Climate Security: The Role of Knowledge and Scientific Information in the Making of a Nexus

This book presents an empirical study of the role of knowledge in the making of the climate-security nexus.

Climate change might give the Soviet Union a competitive advantage in the Cold War. Extreme droughts contributed to wars in Darfur, Syria or Yemen. Melting sea ice creates geopolitical risks. Russia's climate-destroying hydrocarbons enabled its invasion of Ukraine. These are just some of the many ways in which climate change and conflicts have been linked into a climate-security nexus. In this innovative book, Matti Goldberg considers how such connections are constructed and asks to what extent they are driven by evidence and science. Goldberg describes the tools used to present the wars of Darfur and Syria as "climate wars" and considers the fragmented role of the sciences in those presentations as well as the resulting patterns of influence and marginalization of impacted populations. The author also highlights how the international community can better integrate the situations of people at the frontlines of climate change into policymaking and, based on an analysis of the dynamic nature of power, identifies potential entry points for positive change.

This book is a must-read for researchers interested in climate-security links, in science-policy interfaces, and in the formation of nexuses of issues in international politics. It is also of interest to practitioners working on the climate-security nexus and science-policy interfaces.

Matti Goldberg has worked for 14 years at the UN climate change secretariat, supporting the efforts of governments to develop landmark climate policy instruments, such as the Paris Agreement. He currently works as an independent climate policy consultant. He holds a master's degree from the University of Munich. In October 2022, he defended his PhD thesis on the role of knowledge in linking climate change and conflicts at the Technical University of Darmstadt.

Routledge Advances in Climate Change Research

For more information about this series, please visit: www.routledge.com/Routledge-Advances-in-Climate-Change-Research/book-series/RACCR

Climate Security: The Role of Knowledge and Scientific Information in the Making of a Nexus

Matti Goldberg

Routledge
Taylor & Francis Group

LONDON AND NEW YORK

from Routledge

First published 2024
by Routledge
4 Park Square, Milton Park, Abingdon, Oxon OX14 4RN

and by Routledge
605 Third Avenue, New York, NY 10158

Routledge is an imprint of the Taylor & Francis Group, an informa business

British Library Cataloguing-in-Publication Data
A catalogue record for this book is available from the British Library

ISBN: 978-1-032-58793-6 (hbk)
ISBN: 978-1-032-58794-3 (pbk)
ISBN: 978-1-003-45152-5 (ebk)

DOI: 10.4324/9781003451525

Typeset in Times New Roman
by Apex CoVantage, LLC

Other information
Doctoral dissertation at the Technical University of Darmstadt.

Contents

Abbreviations

ANT	Actor-Network Theory
AR5	Fifth Assessment Report of the IPCC
AR6	Sixth Assessment Report of the IPCC
AVHRR	Advanced Very High Resolution Radiometer
BCSC	Berlin Climate and Security Conference
CCS	Center for Climate and Security
CMIP5	Coupled Model Intercomparison Project Phase 5
CNA	Center for Naval Analysis
CNN	Cable News Network
CTEC	Center for Terrorism, Extremism and Counterterrorism
DFID	Department for International Development (United Kingdom)
DoD	United States Department of Defense
DPPA	Department of Political and Peacebuilding Affairs
DWD	Deutscher Wetterdienst
EDI	Effective Drought Index
EU	European Union
FAO	Food and Agriculture Organization of the United Nations
FSA	Free Syrian Army
GBN	Global Business Network
GFFO	German Federal Foreign Office
GHCN	Global Hydrological Climatology Network
GHG	Greenhouse Gas
GIMMS	Global Inventory Modeling and Mapping Studies
GLCF	Global Land Cover Facility
GOS	Government of Sudan
GPCC	Global Precipitation Climatology Center
GPCP	Global Precipitation Climatology Project
GWR	Guinness World Records
HRW	Human Rights Watch
ICARDA	International Center for Agricultural Research in the Dry Areas
ICC	International Criminal Court
ICG	International Crisis Group
IFPRI	International Food Policy Research Institute

ILRI	International Livestock Research Institute
IMCCS	International Military Council on Climate and Security
IPCC	Intergovernmental Panel on Climate Change
IR	International Relations
IRIN	Integrated Regional Information Networks
ISIS	Islamic State in Iraq and Syria
JCPS	Johannesburg Center for Policy Studies
JEM	Justice and Equality Movement
KNMI	Royal Netherlands Meteorological Institute
KMA	Korea Meteorological Administration
LWE	Liquid Water Equivalent
MAAR	Ministry of Agriculture and Agrarian Reform
MSC	Munich Security Conference
NASA	United States National Aeronautics and Space Administration
NATO	North Atlantic Treaty Organization
NCDC	National Climatic Data Center
NDVI	Normalized Difference Vegetation Index
NGO	Nongovernmental Organization
NOAA	United States National Oceanic and Atmospheric Administration
NPR	National Public Radio
NULS	Norwegian University of Life Sciences
NYT	New York Times
ODI	Overseas Development Institute
PIK	Potsdam Institute for Climate Impact Research
PNAS	Proceedings of the National Academy of Sciences
PRIO	Peace Research Institute Oslo
RNC	Republican National Committee
SAIC	Science Applications International Corporation
scPDSI	self-calibrating Palmer Drought Severity Index
SECS	Sudanese Environment Conservation Society
SLM	Sudanese Liberation Movement
SMA	Sudan Meteorological Authority
SPC	State Planning Commission of Syria
SPI	Standard Precipitation Index
STS	Science and Technology Studies
UCDP	Uppsala Conflict Data Program
UEA CRU	University of East Anglia's Climate Research Unit
UNCSM	United Nations Climate Security Mechanism
UNDP	United Nations Development Programme
UNEP	United Nations Environment Programme
UNFCCC	United Nations Framework Convention on Climate Change
UNHCR	United Nations High Commissioner for Refugees
UNOCHA	United Nations Office for the Coordination of Humanitarian Affairs
UNSC	United Nations Security Council
UPAP	University for Peace Africa Programme

USDA	United States Department of Agriculture
WHOI	Woods Hole Oceanographic Institute
WMO	World Meteorological Organization
YCC	YaleClimateConnections

1 Introduction

The climate-security paradox: acceleration in spite of limited evidence

Global policy debates sometimes involve unexpected contributors. The website of Guinness World Records (GWR) declares that the Darfur civil war that began in 2003 was the "first climate war". When, by whom and how this questionable record was set is unclear. Guinness declined to comment. Whatever the background, amid pictures of record-setting situations (including a man cracking watermelons with his head and possibly the largest mouth in the world), the website places Darfur in the category of "firsts" and describes how the lack of rainfall and desertification triggered a resource competition and armed conflicts between Muslims and Christians and led to the "what is widely regarded as the world's first climate change war" (GWR 2019).

Some years earlier, an article in the *Huffington Post* connected climate, the Syrian civil war and the rise of the Islamic State in Iraq and Syria (ISIS):

> Many in the West remain unaware that climate played a significant role in the rise of Syria's extremists. A historic drought afflicted the country from 2006 through 2010, setting off a dire humanitarian crisis for millions of Syrians. Yet the four-year drought evoked little response from Bashar al-Assad's government. Rage at the regime's callousness boiled over in 2011, helping to fuel the popular uprising. In the ensuing chaos, ISIS stole onto the scene.
> (Berkell 2014)

The civil wars in Darfur and Syria are among the deadliest conflicts of the 21st century. But they have also decisively influenced the broader climate-security debate. The Darfur conflict was the first to be labeled a "climate war", and the Syrian crisis remains the most referenced example of climate-conflict links. The two conflicts are the milestones through which the climate-security debate has transitioned from an environmental security issue into a global matter of war and peace.

The two conflicts form a part of a broad debate about environmental security and climate-security links. Environmental security debates began in parallel to the intensification of the global environmental movements in the 1970s. Drawing

DOI: 10.4324/9781003451525-1

heavily on Malthusian assumptions of resource scarcity, the debates were driven in particular by the 1972 Club of Rome's *Limits of Growth* report (Meadows et al. 1972), several influential research papers that proposed broadening the concept of security to encompass environmental issues (e.g. Brown 1977; Ullman 1983; Ehrlich and Ehrlich 1988) and the activities of many environmental nongovernmental organizations (NGOs) (see Diez et al. 2016). Building on the environmental security debate, the climate-security debate began in the context of the post-Cold War reframing of security (see e.g. Buzan et al. 1998; Diez et al. 2016) and was focused in particular on the different ways in which climate change could lead to new types of conflicts (e.g. Homer-Dixon 1991; Myers 1995). Since then, the climate-security debate has expanded to consider a broad range of issues such as migration (Myers 1993), water and energy security, as well as human security, among many other dimensions, and has increasingly informed various national security strategies (Diez et al. 2016).

Recent governance developments indicate that the broader climate-security debate continues to accelerate. For example, on 27 January 2021, US President Biden identified climate change as an "essential element of United States foreign policy and national security", launched a *National Intelligence Estimate* on its security impacts and mandated the consideration of climate risks in defense planning (The White House 2021). In February 2021, the European Union (EU) published its climate adaptation strategy, which emphasized that "climate change multiplies threats to international stability and security" and identified adaptation as a tool for conflict prevention and peacebuilding (EC 2021, 19). On 23 February 2021, the United Nations Security Council (UNSC) met again to debate climate and security. On 7 June 2021, the International Military Council on Climate and Security (IMCCS) launched the *World Climate and Security Report 2021* (IMCCS 2021). And, on 14 June 2021, North Atlantic Treaty Organization (NATO) leaders acknowledged the importance of addressing security impacts of climate change, initiated the integration of climate aspects in operations and set up a *NATO Centre of Excellence on Climate and Security* (NATO 2021). Finally, even as the 2022 Russian invasion of Ukraine has reconfigured the global security discourse, recent publications have begun to connect it with climate change (Sikorsky et al. 2022).[1] These are only a few examples of how climate has been integrated into Euro-Atlantic security discourses. They show how two hegemonial narratives of contemporary politics – climate and security – are merged into a "climate-security nexus" in national and international policy forums, military plans, research and institutions.

For many, the debate on whether climate and security are connected has concluded with a yes. However, not everyone agrees. Russia and China oppose decision-making on climate in the Security Council (UNSC 2018, 2019). Researchers have criticized the climate-security debate as simplistic, as disconnected from facts, as a get-out-of-jail-free card for oppressive regimes and alarmist, and for its tendency to militarize environmental questions (see Barnett 2001; De Waal 2007; Kevane and Gray 2008; Verhoeven 2011; Selby and Hoffmann 2014; Selby et al. 2017, 2022).[2] Parallel to the broader climate-security debate, scholars have questioned in particular the knowledge basis of connecting climate change and specific conflicts (Selby and Hoffmann 2014; Fröhlich 2016; Selby et al. 2017).

Social scientific research has highlighted the role of knowledge in securitization of environmental issues (Buzan et al. 1998; Berling 2011). This is particularly true for climate change because understanding it requires scientific tools (see Edwards 2013; IPCC 2018). However, reviews of evidence for climate-conflict links have brought ambiguous results (Scheffran et al. 2012; Theisen et al. 2013; IPCC 2014; Detges 2017; Hardt and Scheffran 2019; Mach et al. 2019).

This indicates a contradictory logic: (a) climate-security links are expected to rely on scientific evidence; (b) however, the evidence is inconclusive and contested; (c) but the development of the climate-security nexus continues to intensify. In other words, securitization of climate change – considered a science-driven issue – is advancing although the science remains disputed. This leads to questions about the degree to which the climate-security nexus is science-driven, and the exact role of forms of knowledge in its development. Possibly, the combination of ambiguous evidence and a solidifying climate-security nexus indicates a disconnection between science and policymaking.

To better understand this contradiction, this book presents a study of one aspect of climate-security debates: how the climate-security nexus is being advanced by connecting climate change and specific conflicts through the use of knowledge resources. It describes how actors articulate climate-conflict links, what knowledge resources they use to enable meaningful statements about those links, how the debate allows connecting the traditionally separate domains of "climate" and "security", as well as the consequences of such connecting. An important distinction is that this book does not provide a comprehensive study of the broader climate-security debate described in this section. Instead, its focus on the making of the two "climate wars" of Darfur and Syria should help understand better what enabled GWR to label Darfur as the "first climate war", the *Huffington Post* to connect climate change and ISIS, and western policy actors to converge around climate-security.

Climate-security: one nexus among many

"Climate" and "security" were, until the 2000s, mostly disconnected and characterized by their own practices, discourses and institutions. However, efforts to merge the two into "climate-security"[3] have intensified, and the institutionalization of the nexus has accelerated since the mid-2010s.[4] This is significant because climate and security are central issues in international politics. They both permeate policy discourses and national priority lists, influence other issues and involve calls for extraordinary measures. In the climate-security nexus, two of today's hegemonial governance areas are being merged.

However, the climate-security nexus is only one of the nexuses that populate international policy discussions. Stipulations of nexuses between governance areas are common in development discourses (see Stern and Öjendal 2010; Cairns and Krzywoszynska 2016; Carling 2017; Liu et al. 2018; Williams et al. 2018). In IR research, Ernst Haas has considered, in context of interdependence and regime formation, how issues are linked into packages of "issue areas" (1980, 361). He described the emergence of an international monetary policy and the oceans as issue areas in the 1940s and 1960s as policymakers recognized the interdependencies between their components (Ibid., 364–6). Stern and Öjendal (2010) analyzed the "development-security nexus", observing that, while nexuses permeate development discourses, they mean different things to different actors and are open to all kinds of uses, including problematic ones. This study complements studies of issue-linkages and nexuses by considering how the climate-security nexus is being formed through knowledge practices involved in the making of "climate wars". It focuses on the initial stages of the climate-security debates prior to institutionalization, established practices or regime formation, thus considering the climate-security nexus "in the making" (Latour 2005, 118).

Knowledge and international politics

Knowledge is pertinent for the climate-security nexus because we understand the climate through scientific measurements and presentations (see Edwards 2013; IPCC 2018). Climate-security debates are permeated by references to scientific studies and expert statements. Haas observed the central role of knowledge and expertise in the establishment of issue-areas (1980), highlighting that knowledge is the "basic ingredient for exploring the development of issue-areas" in international politics (1980, 367), that consensual knowledge relevant to policy goals can lead to packages of issue-areas (Ibid., 372) and that the formation of issue-areas is often preceded by consensus about intellectual strategies or causal understandings (Ibid., 374). He also emphasized that knowledge is not sufficient for the formation of issue packages because national interests will not change just because of new insights about issues, and highlighted the chaotic nature of the use of knowledge, arguing that often, when linking issues, "causal sequences are assumed or guessed at rather than studied fully" (Ibid., 378).

Recently, securitization research has noted the role of sciences in environmental security (Buzan et al. 1998; Rychnovská et al. 2017). Some have described that role systematically (Berling 2011; Rothe 2017), while others have focused on knowledge, practices and materials in security discourses (Aradau 2010; Huysmans 2011; Mayer 2012; Rothe 2017; Rychnovská et al. 2017; Salter 2019). Similarly, practice-oriented perspectives have considered how knowledge of issues is embedded in practices, how those practices reify the phenomena of IR and evolve into structures and how the practices transform knowledge (see Lederer 2012; Bueger and Gadinger 2018). This study complements this diverse landscape by considering how climate- and security-related knowledge are mobilized by actors in their efforts to link climate change with the Darfur and Syria conflicts.

How to study knowledge and nexus formation?

But what research strategy would help understand nexus formation and knowledge? IR research has generated insights about the role of knowledge in policymaking, discourses of security and established practices. However, to understand the role of knowledge in the making of climate wars and the formation of the climate-security nexus – an emerging issue without stable institutions or practices – this study builds on science and technology studies (STS) and in particular actor-network theory (ANT) to adopt an approach referred to as a *sociology of translations*.[5] ANT-based studies focus on networks to understand how social relations and arrangements develop through the association of human and nonhuman actors into actor-networks through processes of *translation* (Callon 1986; Latour 2005; Irwin 2008).

Several studies have discussed how STS and ANT approaches can help study knowledge in IR and security studies (Barry 2013; Best and Walters 2013(a), 2013(b); Allan 2017; Bueger and Stockbruegger 2017; Braun et al. 2019; Salter 2019). Such approaches have been applied, for example, to international security networks (Aradau 2010), global practices and enabling knowledge networks of torture (Austin 2015, 2016), as well as environmental and climate security (Mayer 2012; Rothe 2017).

Building on such perspectives, this book is about how actors associate or, in methodological terms, "translate" climate and the civil wars of Darfur and Syria together. It focuses on how such associations are created with various knowledge resources, in particular scientific information and methods (e.g. publications, statistical tools and indices). Based on an empirical mapping of those resources, this book considers three broad aspects of knowledge and climate-conflict links:

- How to characterize the role of knowledge resources in the making of the climate-security nexus – in other words, how science-driven are efforts to securitize climate change?
- How do imbalances in availability of knowledge in the making of climate wars illustrate epistemic "blind spots" of international governance, reproduce marginalization and lead to confirmation biases?
- How are patterns of power and influence intertwined with the knowledge resources mobilized during this part of nexus formation?

In addition, this book suggests how policymakers, practitioners and researchers can strive toward a better balance between evidence and narratives.

This book is not about whether conflicts are connected with climate change, whether the evidence for climate-conflict links is "good" or "bad", the motivations of policymakers to build links or the speech acts connecting climate and conflicts. Many studies have covered such questions. Rather, it considers how knowledge features, in all its diversity and unpredictability, and beyond conventional models of the science-policy relationship, in the making of climate wars and thus the climate-security nexus, and with what consequences.

Through this approach, this book is intended as a contribution to climate-security research by describing the role of knowledge resources in climate-conflict debates. It also provides an impulse toward the empirical study of nexuses in international relations (IR) by looking at the dynamics and consequences of the ways in which two hegemonial issues of our time – climate and security – are being merged by connecting specific wars with climate change. In addition, by focusing on the role of knowledge in policymaking, this book contributes to the conversation on the relationship between knowledge and power going back in the Western intellectual tradition as far as Plato. Finally, through its research strategy, this book is intended to inspire scholars seeking entry points for the study of knowledge in nexus formation and in other policy processes.

Structure of this book

This book is organized in nine chapters. The first three chapters present the phenomenon of climate-security, how it relates to nexus formation and the research strategy of this book. Chapter 2 describes how notions of security have broadened to encompass environmental and other aspects, how researchers have tried to understand climate-security links and how those links have gained political and institutional traction. It outlines the controversies about climate-security links and highlights the contradictory coexistence of science-drivenness of climate change, disputed evidence and accelerating securitization.

Chapter 3 introduces the trend of nexuses in international politics, outlining patterns of nexus formation, along with how IR and development studies have focused on issue-linkages, nexuses as a governance approach and critical perspectives on nexuses. The chapter also describes how the study of nexuses can benefit from empirical studies of their formation and from consideration of the role of knowledge, in particular, when it comes to environmental issues, which involve significant knowledge components.

Chapter 4 presents the research strategy of this book. It outlines how this book approaches a key aspect of the formation of the climate-security nexus by considering how climate and the conflicts of Darfur and Syria are being translated together by actors. It highlights how the strategy, inspired by ANT, is to follow the activities of actors to connect climate and the conflicts, and to characterize the network of associations resulting from those translations, while avoiding a priori assumptions. Here, the analytical concepts and principles of the strategy, and the rationale for choosing it, are elaborated. The chapter outlines three analytical steps toward understanding knowledge and climate-conflict links: (1) mapping how actors translate climate and conflicts together; (2) describing the knowledge resources they mobilize; and (3) characterizing the resulting actor-networks.

Chapters 5 and 6 describe how climate and the two paradigmatic "climate wars" of Darfur and Syria have been linked through translations and knowledge resources. Chapter 5 focuses on how actors do this in the context of the Darfur civil war. It describes the actors, the controversies they debated, their arguments and the knowledge resources they mobilized. The chapter identifies how actors consider controversies about the status of Darfur as the "first climate war", lack of rains, resource scarcity, farmer-pastoralist conflicts, implications of climate-Darfur links beyond the conflict itself and whether climate-Darfur narratives enable Khartoum to avoid responsibility for the conflict. Chapter 6 describes how actors associated climate and the Syrian civil war, with a similar focus on the actors, controversies, translations and knowledge resources. In this case, actors debate the severity of Syria's pre-war drought, its links with climate change, the migration impacts of the drought, the role of rural migrants in Syria's 2011 uprisings and a possible link between climate change and ISIS.

Chapter 7 considers basic characteristics of the climate-conflict actor-networks emerging from the translations described in Chapters 5 and 6, including key similarities and differences between the two debates. It highlights how linear causal mechanisms articulated by actors at early stages of the debate provide the basis for connecting climate and the two conflicts, and how those mechanisms become the structure of the debates that follow. The chapter then considers how both climate-conflict debates progress through three overlapping stages characterized by changing modalities of translation: initial articulations of the linear causal mechanisms, science-based reviews of the mechanism components and efforts to review and synthesize existing evidence. Finally, the chapter describes the diversity of actors, knowledge resources and the channels of circulation of translations involved in the debates.

Chapter 8 considers the role of knowledge resources, in particular scientific knowledge, in climate-conflict actor-networks in light of the linear causal mechanisms, modalities of translation and the diversity of actors, resources and circulation channels. It thereby sheds light on the contradictory co-existence of disputed evidence and accelerating securitization of climate change. The chapter describes an epistemic landscape that bears little resemblance to idealized linear science-policy relations: knowledge resources are heterogenous and fragmented; the sciences work in reaction to narratives; and science-policy boundaries are regularly blurred. It also highlights how the scientific basis of the linear causal mechanisms is imbalanced toward the natural sciences, and involves limited resources for understanding human aspects, leading to marginalization of impacted populations, stereotypical presentations and risks of confirmation bias in climate-conflict research. Finally, it considers how the dynamics of knowledge lead to changing patterns of influence within the climate-conflict actor-networks, describing how influence, rather than being an attribute of actors, results from changing constellations of knowledge resources and changing modalities of translations.

Finally, Chapter 9 recaps the central findings, highlights key challenges related to knowledge, climate wars and nexus formation in a post-truth era, and discusses possible solutions and further research areas. It considers how the fragmentary integration of the sciences in climate-conflict debates and the focus on narrative coherence contradicts global aspirations toward science-based policies, incentivizing marginalization, suboptimal policymaking and possibly climate skepticism. However, the chapter argues that given the dynamic nature of influence, knowledge arrangements, institutions and rules of engagement are not cast in stone but, with balanced adjustments, provide multiple channels toward change. The chapter discusses how change could be pursued by designing climate-security research projects in ways that integrates multiple perspectives. It would also be important to find ways to better use up-to-date information from the front lines of conflicts, while ensuring its quality in climate-security research. In addition, existing institutions could develop efforts to correct information imbalances in climate-security research, so that the entire continuum of questions from environmental change to social impacts would be more adequately covered. Such changes could help articulate scientifically informed nexus policies. Finally, the chapter suggests potential future research areas and concluding observations are highlighted.

Notes

1 These connections have been made by describing the two as part of a cluster of interconnected risks. This includes the notion that Russian attacks on nuclear sites might slow the development of nuclear energy, which would reduce the effectiveness of mitigation and worsen climate-related security risks (Parthemore and Rezzonico 2022). Another perspective is that (a) the war underlines the need to accelerate clean energy development; (b) the "ecological security" of Ukraine is degrading; (c) the crisis impacts global food security, and (d) because of Russia's own "climate-security vulnerabilities" (Sikorsky et al. 2022).
2 Criticism also comes from conservative media (Moran 2020) and policymakers, including presidential candidate Jeb Bush, Senator Marco Rubio and RNC Chairman Reince Priebus, who argue that climate-security debates distract from "real" national security issues.
3 The term "climate-security" was first used by UK Foreign Minister Beckett in her speech in Berlin (Beckett 2006).
4 Including through the establishment of the UNCSM, UNSC debates, the Group of Friends of Climate and Security, and many non-governmental activities.
5 The term "sociology of translations" was originally used by Michel Callon in his study on the relationships between researchers, scallops and fishermen on St. Brieuc Bay (1986).

Literature

Allan, B.B. 2017: *From subjects to objects: Knowledge in international relations theory*. In: European Journal of International Relations, 24(4), 841–64.

Aradau, C. 2010: *Security that matters: Critical infrastructure and objects of protection*. In: Security Dialogue, 41(5), 491–514.

Austin, J.L. 2015: *We have never been civilized: Torture and the materiality of world political binaries*. In: European Journal of International Relations, 23(1), 49–73.

Austin, J.L. 2016: *Torture and the material-semiotic networks of violence across borders*. In: International Political Sociology, 10, 3–21.

Barnett, J. 2001: *Climate Change and Security*. Tyndall Center for Climate Change Research, Working Paper 7, Manchester. October.

Barry, A. 2013: *The translation zone: Between actor-network theory and international relations*. In: Millennium: Journal of International Studies, 41(3), 413–29.

Beckett, M. 2006: *Foreign Policy and Climate Security*. Speech on 24 October, Berlin. At: https://webarchive.nationalarchives.gov.uk/20061102073121/www.fco.gov.uk/servlet/Front/TextOnly?pagename=OpenMarket/Xcelerate/ShowPage&c=Page&cid=10070293 91647&to=true&a=KArticle&aid=1161588023142 (accessed 28 September 2009).

Berkell, K.A. 2014: *How climate change helped ISIS*. Huffington Post. 29 November. At: www.huffpost.com/entry/how-climate-change-helped_b_5903170 (accessed 20 February 2022).

Berling, T.V. 2011: *Science and securitization: objectivation, the authority of the speaker and mobilization of scientific facts*. In: Security Dialogue, 42(4–5), 385–97.

Best, J. and Walters, W. 2013(a): *Actor-network theory and international relationality: Lost (and found) in translation.* In: International Political Sociology, 7(3), 332–4.

Best, J. and Walters, W. 2013(b): *Translating the sociology of translations.* In: International Political Sociology, 7(3), 345–9.

Braun, B.; Schindler, S. and Wille, T. 2019: *Rethinking agency in international relations: performativity, performances and actor-networks.* In: Journal of International Relations and Development, 22, 787–807.

Brown, L.R. 1977: *Redefining National Security.* Worldwatch Paper 14. Worldwatch Institute, Washington, DC.

Bueger, C. and Gadinger, F. 2018: *International Practice Theory.* Palgrave Macmillan, Cham.

Bueger, C. and Stockbruegger, J. 2017: *Actor-network theory: objects and actants, networks and narratives.* In: McCarthy, D. (Ed.): Technology and World Politics. Routledge, Abingdon, 42–59.

Buzan, B.; Waever, O. and de Wilde, J. 1998: *Security – A New Framework for Analysis.* Lynne Rienner Publishers Inc., London.

Cairns, R. and Krzywoszynska, A.D. 2016: *Anatomy of a buzzword: The emergence of "the water-energy-food nexus" in UK natural resource debates.* In: Environmental Science and Policy, 64, 164–70.

Callon, M. 1986: *Some elements of a sociology of translation: domestication of the scallops and the fishermen of St. Brieuc Bay.* In: Law, J. (Ed.): Power, Action and Belief: A New Sociology of Knowledge? Routledge, London, 196–223.

Carling, J. 2017: *Thirty-Six Migration Nexuses, and Counting.* At: https://jorgencarling.org/2017/07/31/thirty-six-migration-nexuses-and-counting/ (accessed 9 September 2020).

De Waal, A. 2007: *Is climate change the culprit for Darfur?* African Arguments. 25 June. At: https://africanarguments.org/2007/06/25/is-climate-change-the-culprit-for-darfur/ (accessed 6 November 2020).

Detges, A. 2017: *Climate and Conflict: Reviewing the Statistical Evidence – a Summary for Policymakers.* Adelphi, Berlin.

Diez, T.; von Lucke, F. and Wellmann, Z. 2016: *The Securitisation of Climate Change: Actors, Processes and Consequences.* Routledge Prio New Security Studies. Routledge, Oxon.

Edwards, P.N. 2013: *A Vast Machine – Computer Models, Climate Data, and the Politics of Global Warming.* MIT Press, Cambridge, MA.

Ehrlich, P.R. and Ehrlich, A.H. 1988: *The Environmental Dimensions of National Security.* Stanford Institute for Population and Resource Studies, Stanford, CA.

European Commission (EC) 2021: *Forging a Climate-Resilient Europe – the New EU Strategy on Adaptation to Climate Change.* 24 February. At: https://ec.europa.eu/clima/sites/clima/files/adaptation/what/docs/eu_strategy_2021.pdf (accessed 29 May 2021).

Fröhlich, C. 2016: *Climate migrants as protestors? Dispelling misconceptions about global environmental change in pre-revolutionary Syria.* In: Contemporary Levant, 1(1), 38–50.

GWR 2019: *First Climate Change War.* At: www.guinnessworldrecords.com/world-records/first-climate-change-war (accessed 7 September 2021).

Haas, E.B. 1980: *Why collaborate?: Issue-linkage and international regimes.* In: World Politics, 32(3), 357–405.

Hackett, E.J.; Amsterdamska, O.; Lynch, M. and Wajcman, J. (Eds.) 2008: *The Handbook of Science and Technology Studies.* 3rd Edition. MIT Press, Cambridge, MA.

Hardt, J.N. and Scheffran, J. 2019: *Environmental Peacebuilding and Climate Change: Peace and Conflict Studies at the Edge of Transformation.* Toda Peace Institute Policy Brief No. 68. December. At: https://toda.org/policy-briefs-and-resources/policy-briefs/

environmental-peacebuilding-and-climate-change-peace-and-conflict-studies-at-the-edge-of-transformation.html (accessed 9 January 2020).

Homer-Dixon, T. 1991: *On the threshold: environmental changes as causes of acute conflict.* In: International Security, 16(2), 76–116.

Huysmans, J. 2011: *What's in an act? On security speech acts and little security nothings.* In: Security Dialogue, 42(4–5), 371–83.

International Military Council on Climate and Security 2021: *World Climate and Security Report 2021.* Center for Climate and Security. At: https://imccs.org/wp-content/uploads/2021/06/World-Climate-and-Security-Report-2021.pdf (accessed 6 August 2021).

IPCC 2014 (Adger, W.N.; Pulhin, J.M.; Barnett, J.; Dabelko, G.D.; Hovelsrud, G.K.; Levy, M.; Oswald Spring, Ú. and Vogel, C.H.): *Human security.* In: Field, C.B.; Barros, V.R.; Dokken, D.J.; Mach, K.J.; Mastrandrea, M.D.; Bilir, T.E.; Chatterjee, M.; Ebi, K.L.; Estrada, Y.O.; Genova, R.C.; Girma, B.; Kissel, E.S.; Levy, A.N.; MacCracken, S.; Mastrandrea, P.R. and White, L.L. (Eds.): Climate Change 2014: Impacts, Adaptation, and Vulnerability. Part A: Global and Sectoral Aspects. Contribution of Working Group II to the Fifth Assessment Report of the IPCC. Cambridge University Press, Cambridge and New York, NY, 755–91.

IPCC 2018 (Matthews, J.B.R. (Ed.)): *Annex I: Glossary.* In: Masson-Delmotte, V.; Zhai, P.; Pörtner, H.-O.; Roberts, D.; Skea, J.; Shukla, P.R.; Pirani, A.; Moufouma-Okia, W.; Péan, C.; Pidcock, R.; Connors, S.; Matthews, J.B.R.; Chen, Y.; Zhou, X.; Gomis, M.I.; Lonnoy, E.; Maycock, T.; Tignor, M. and Waterfield, T. (Eds.): Global Warming of 1.5°C: An IPCC Special Report on the Impacts of Global Warming of 1.5°C Above Pre-Industrial Levels and Related Global Greenhouse Gas Emission Pathways, in the Context of Strengthening the Global Response to the Threat of Climate Change, Sustainable Development, and Efforts to Eradicate Poverty, 539–562. At: www.ipcc.ch/site/assets/uploads/sites/2/2019/06/SR15_AnnexI_Glossary.pdf (accessed 5 July 2021).

Irwin, A. 2008: *STS perspectives on scientific governance.* In: Hackett et al. 2008, 583–608.

Kevane, M. and Gray, L. 2008: *Darfur: Rainfall and conflict.* In: Environmental Research Letters, 3. At: https://iopscience.iop.org/article/10.1088/1748-9326/3/3/034006/pdf (accessed 21 November 2019).

Latour, B. 2005: *Reassembling the Social – an Introduction to Actor-Network Theory.* Oxford University Press, Oxford.

Lederer, M. 2012: *The practice of carbon markets.* In: Environmental Politics, 21(4), 640–56.

Liu, J.; Hull, V.; Godfray, C.; Tilman, D.; Gleick, P.; Hoff, H.; Pahl-Wostl, C.; Xu, Z.; Sun, J. and Li, S. 2018: *Nexus approaches to global sustainable development.* In: Nature Sustainability, 1, 466–76.

Mach, K.J.; Kraan, C.M.; Adger, W.N.; Buhaug, H.; Burke, M.; Fearon, J.D.; Field, C.B.; Hendrix, C.S.; Maystadt, J.F.; O'Loughlin, J.; Roessler, P.; Scheffran, J.; Schultz, K.A. and von Uexkull, N. 2019: *Climate as a risk factor for armed conflict.* In: Nature, 571, 193–7.

Mayer, M. 2012: *Chaotic climate change and security.* In: International Political Sociology, 6, 165–85.

Meadows, D.H.; Meadows, D.L.; Randers, J. and Behrens, W.W. III 1972: *The Limits to Growth: A Report of the Club of Rome's Project on the Predicament of Mankind.* Universe Books, New York.

Moran, S. 2020: *Fact Check: Joe Biden Says Climate Change Caused Darfur Conflict.* Breitbart News. 15 March. At: www.breitbart.com/politics/2020/03/15/fact-check-joe-biden-says-climate-change-caused-darfur-conflict/ (accessed on 6 November 2020).

Myers, N. 1993: *Environmental refugees in a globally warmed world.* In: BioScience, 43(11), 752–61.

Myers, N. 1995: *Environmental Exodus: An Emergent Crisis in the Global Arena*. The Climate Institute, Washington, DC.

NATO 2021: *Press Conference by NATO Secretary General Jens Stoltenberg Following the Meeting of NATO Heads of State and Government*. 14 June. At: www.nato.int/cps/en/natohq/opinions_184959.htm (accessed 23 June 2021).

Parthemore, C. and Rezzonico, A. 2022: *Nuclear Threats in Ukraine Today and Their Implications for Global Security Tomorrow*. At: https://councilonstrategicrisks.org/2022/03/15/nuclear-threats-in-ukraine-today-and-their-implications-for-global-security-tomorrow/ (accessed 4 August 2022).

Rothe, D. 2017: *Seeing like a satellite: Remote sensing and the ontological politics of environmental security*. In: Security Dialogue, 48(4), 334–53.

Rychnovská, D.; Pasgaard, M. and Berling, T.V. 2017: *Science and security expertise: Authority, knowledge, subjectivity*. In: Geoforum, 84, 327–31.

Salter, M. 2019: *Security actor-network theory: Revitalizing securitization theory with Bruno Latour*. In: Polity, 51(2), 349–3,64.

Scheffran, J.; Brzoska, M.; Kominek, J.; Lin, P.M. and Schilling, J. 2012: *Climate change and violent conflict*. In: Science, 336, 869–71.

Selby, J.; Dahi, O.; Fröhlich, C. and Hulme, M. 2017: *Climate change and the Syrian civil war revisited*. In: Political Geography, 60(1), 232–44.

Selby, J.; Daoust, G. and Hoffmann, C. 2022: *Divided Environments: An International Political Ecology of Climate Change, Water and Security*. Cambridge University Press, Cambridge.

Selby, J. and Hoffmann, C. 2014: *Beyond scarcity: Rethinking water, climate change and conflict in the Sudans*. In: Global Environmental Change, 29, 360–70.

Sikorsky, E.; Barron, E. and Hugh, B. 2022: *Climate, Ecological Security and the Ukraine Crisis: Four Issues to Consider*. CCS Briefer No. 31. 12 March. At: https://climateandsecurity.org/wp-content/uploads/2022/03/Climate-Ecological-Security-and-the-Ukraine-Crisis_Four-Issues-to-Consider_BRIEFER-31_2022_11_3.pdf (accessed 21 July 2022).

Stern, M. and Öjendal, J. 2010: *Mapping the security development nexus: Conflict, complexity, cacophony, convergence?* In: Security Dialogue, 41(1), 5–30.

Theisen, O.M.; Gleditsch, N.P.; Buhaug, H. 2013: *Is climate change a driver of armed conflict?* In: Climatic Change, 117(3), 613–25.

Ullman, R.H. 1983: *Redefining security*. In: International Security, 8(1), 129–53.

UNSC 2018: *Addressing Security Council, Pacific Island President Calls Climate Change Defining Issue of Next Century, Calls for Special Representative on Issue*. UN Press Release SC/13417 on the 8307th Meeting of the Security Council. At: www.un.org/press/en/2018/sc13417.doc.htm (accessed 30 May 2015).

UNSC 2019: *Massive Displacement, Greater Competition for Scarce Resources Cited as Major Risks in Security Council Debate on Climate-Related Threats*. UN Meetings Coverage. 8451st Meeting of the Security Council. 25 January. At: www.un.org/press/en/2019/sc13677.doc.htm (accessed 7 January 2020).

Verhoeven, H. 2011: *Climate change, conflict and development in Sudan: Global neo-Malthusian narratives and local power struggles*. In: Development and Change, 42(3), 679–707.

The White House 2021: *Executive Order on Tackling the Climate Crisis at Home and Abroad*. 27 January. At: www.whitehouse.gov/briefing-room/presidential-actions/2021/01/27/executive-order-on-tackling-the-climate-crisis-at-home-and-abroad/ (accessed 29 May 2021).

Williams, J.; Bouzarovski, S. and Swyngedouw, E. 2018: *The urban resource nexus: On the politics of relationality, water – energy infrastructure and the fallacy of integration*. In: Politics and Space, 37(4), 652–69.

2 The climate-security contradiction

Accelerating securitization under disputed evidence

Climate and security: toward merging two hegemonial domains

The climate-security nexus is particularly interesting from an international policy perspective because it is an effort to merge two well-established and hegemonial governance areas. The areas of climate and security dominate global policy conversations, strongly influence other governance areas and involve calls for extraordinary responses to existential threats.

However, the two have traditionally been separated. At the UNSC debate on 23 February 2021, UK prime minister Johnson challenged the common prejudice of considering climate to be "green stuff from a bunch of tree-hugging tofu munchers" (UNSC 2021). Climate is often seen as a "soft" environmental issue, while security issues are allocated to "high politics". This separation results in particular from distinct history, knowledge basis, definitions, conceptualizations, perceptions and governance of the two areas. Climate change, a comparably recent phenomenon of international governance, is defined as

> a change in the state of the climate that can be identified (e.g., by using statistical tests) by changes in the mean and/or the variability of its properties and that persist for an extended period, typically decades or longer.
>
> (IPCC 2018, 544)

While scientific definitions of climate change acknowledge multiple drivers, anthropogenic changes in GHG emissions and land use are considered to be the main driver. Reflecting this, the policy definition of climate change, captured in Article 1 of the United Nations Framework Convention on Climate Change (UNFCCC), is "a change of climate which is attributed directly or indirectly to human activity that alters the composition of the global atmosphere and which is in addition to natural climate variability observed over comparable time periods" (UN 1992). This way, climate change is a process of environmental change driven by anthropogenic GHG emissions, documented by scientific study, which the international community, governments, and civil society recognize as a challenge that needs to be tackled, and for which they have developed policies to reduce emissions and to adapt to climate hazards. As a result, today "people, states, corporations and

DOI: 10.4324/9781003451525-2

international organizations . . . operate giant monitoring and regulatory systems in concerted attempts to change (or preserve) the chemical composition of the global atmosphere" (Corry 2014, 219).

Security, on the other hand, is often understood in IR research and international politics as being "about survival" (Buzan et al. 1998). It is evoked when "an issue is presented as posing an existential threat to a designated referent object", paving the way for extraordinary measures to protect the object (Ibid., 21). While climate change operates in the long term, security issues tend to urgency and immediacy. While "traditional" understandings of security have considered the state, territory and society as referent objects, the concept has expanded to encompass, that is, "human security", "environmental security" and "economic security" (Ibid.). Institutions working on security tend to be national institutions with monopolies of force (e.g. militaries, intelligence services and police), international alliances (e.g. NATO or Collective Security Treaty Organization) and international organizations (e.g. the UNSC, African Union and Shanghai Cooperation Organization).

While both climate change and security are much more complex, these dimensions illustrate the differences between the two. But the two areas are now being increasingly linked in research, policymaking and the practices of international institutions.

Changing notions of security

Until the 1970s, the concept of security used to be strictly about military confrontation, nuclear war and state security. But the emerging economic and environmental agendas, along with the end of the Cold War, triggered an extensive reconsideration of what qualifies as security issues (Buzan et al. 1998; Diez et al. 2016). This reconsideration has largely been a process of what Buzan et al. called "issue-driven widening", where researchers, policymakers and others seek to include new issue areas within the scope of security (Ibid., 2). This pertains in particular to the environment and human security (Barnett 2001). For instance, Falk (1971), Brown (1977) and Ullman (1983), among many others, identified climate change, resource scarcity and ecological degradation as new threats to security and argued that security cannot be confined to military matters. Throughout the debate, authors have highlighted specific expressions of environment-related threats, including the possibilities of environmental change as a driver of conflicts (Homer-Dixon 1991), potential security threats caused by climate-induced migration (Myers 1993, 1995) and Arctic confrontations triggered by melting of the sea ice (Gerhardt et al. 2010). The human security perspective was introduced by UNDP (1994), which suggested reinterpreting security as "human security" with seven elements – economic, food, health, environment, personal, community and political securities – and has expanded into a significant policy and research agenda (UNDP 1994). The diversification of the concept of security was illustrated in Buzan et al.'s landmark *Security: A New Framework for Analysis*, which described how contemporary security discourses involved efforts to apply the notion of security to encompass military, environmental, economic, societal and political sectors. This expansion of the concept of security has enabled the climate-security to develop into its current state as a major area of research and policymaking.

Research and climate-security links: many perspectives, limited consensus

Since the beginnings of the climate-security debate, related research has focused, in particular, on:

- Quantitative studies of correlations between climate change and conflict;
- Studies to understand the role of associated variables in the climate-security link;
- Case studies of specific conflicts;
- Efforts to review the evidence and state of research;
- Qualitative and social scientific, including IR-based, studies on the climate-security nexus.

Quantitative studies

Most climate-security research has focused on correlations between climate parameters and conflict occurrence, as well as quantified projections about the future climate-conflict links. However, studies have found evidence both for and against climate-conflict links. Temperature-conflict correlations were studied by, for example, Zhang et al. (2007), who found that cycles of conflict in 1400–1900 correlated with lower temperatures. In contrast, Burke et al. (2009) noted that temperature increase has led to significant increases in civil war occurrence in Africa. Others have expanded the number of variables, including Hsiang et al. (2013), who found that precipitation and temperature deviations increased conflict risks. In terms of quantified projections, Myers (1993) estimated that climate impacts might force 150 million people to migrate by 2050, Burke et al. (2009) calculated that current climate trends can translate to a 54 per cent increase in conflict occurrence and 393,000 additional battle deaths by 2030, and Hsiang et al. (2013) estimated that each 1°C increase in global temperatures and other climate parameters increases the frequency of intergroup conflict by 14 per cent.

However, others have questioned these correlations. For example, Buhaug observed a negative correlation across Africa between higher temperature and lower precipitation on the one hand, and conflict incidence and battle deaths on the other, concluding that "climate variability is a poor predictor of armed conflict", and emphasizing the role of political, economic and geopolitical factors (2010, 16477). These differences resulted in detailed methodological exchanges between groups of authors, particularly with Hsiang et al. (2013), Hsiang and Meng (2014) and Hsiang and Burke (2014) arguing for climate-conflict correlations, and Buhaug (2010) and Buhaug et al. (2014) against.

Intervening variables

Other studies have focused on how other factors can appear as intervening variables in climate-conflict links. For example, Fjelde and von Uexkull (2012) noted that areas in Africa that have politically excluded minorities have a higher risk of communal conflict after extreme rainfall deviations, and Schleussner et al. (2016) observed that violent conflicts after climate hazards are more likely in ethnically divided societies.

Several studies have looked at climate-conflict links in agriculture-dependent societies. For example, Von Uexkull (2014), Salehyan and Hendrix (2014) and Von Uexkull (2016) found that rainfall-dependent regions in Asia and/or Africa are more likely to see conflicts and unrest after droughts. Maystadt and Ecker (2014) observed that in Somalia, drought can force herders to sell their animals, leading to income loss and incentivizing them to join armed groups. In Indonesia, Caruso et al. (2014) observed that high temperatures can lead loss of rice harvest, lower incomes and a higher risk of violence. In contrast, Buhaug (2015) argued that reduced agricultural outputs do not necessarily lead to unrest.

In addition, the role of political systems in climate-conflict risks has been studied, for example, by Couttenier and Soubeyran (2014), who considered that democratic institutions can mitigate climate-related security risks. Linke et al. (2015) sought to demonstrate that people without conflict mitigation mechanisms tend more toward political violence in times of climatic stress. However, Ide (2017) has emphasized that political issues are difficult to quantify. In terms of infrastructure, Detges (2016) observed that regions in Africa with poor infrastructure are more prone to drought-induced conflict escalation.

Case studies of specific conflicts

As part of the broader climate-security debate, many studies have considered the link between climate and specific conflicts. For instance, in the context of Darfur, Kevane and Gray (2008) criticized the argument that the conflict was aggravated by failed rains by demonstrating that rains did not fail before the conflict. Similarly, Brown (2010) applied satellite data and a vegetation index to conclude that there was no resource scarcity prior to the conflict. Verhoeven (2011) criticized how presenting Darfur in neo-Malthusian terms enabled the government in Khartoum to deny responsibility, and Selby and Hoffmann (2014) and Selby et al. (2022) questioned the credibility of farmer-pastoralist tensions as a factor, given the diversity of Darfurian livelihoods. In 2014, the Intergovernmental Panel on Climate Change's (IPCC's) fifth synthesis report concluded that the evidence for climate-Darfur links is weak.

On Syria, Gleick (2014), Werrell et al. (2015) and Kelley et al. (2015), among others, have argued that a climate-induced drought triggered an agricultural collapse and migration, which destabilized Syrian cities. In contrast, De Châtel (2014) emphasized other factors, Fröhlich (2016) interviewed Syrian refugees to demonstrate that climate-migrants had little to do with the uprisings, and Selby et al. (2017(a), 2017 (b)) questioned the attribution of the drought to climate change and estimated levels of internal migration. As in the case of Darfur, the IPCC concluded in 2022 that the conflict would most likely have happened without climate change.

Climate-conflict links in context of Lake Chad tensions have been studied by Okpara et al. (2016), who observed that water-related conflicts depend on changes in both the environment and in levels of vulnerability, and that both need to be understood to understand conflict dynamics, while Daoust and Selby (2022) questioned the quality of evidence in the climate – Lake Chad discourse.

In the Arctic context, Gerhardt et al. discussed questions of sovereignty of the Northwest Passage, territorial control and options for multilateral governance. They highlighted that Arctic change, while creating risks, also opens opportunities for cooperation and for rethinking the state-based system of "mutually exclusive territorial policies" (2010, 8).

Reviews of evidence

In a review of existing studies, Scheffran et al. (2012) concluded that quantitative studies provide mixed results about climate-conflict links. Theisen et al. (2013) also found studies inconclusive and weak in terms of evidence. Similarly, the fifth assessment report of the IPCC (AR5) concluded that, while climate change will drive insecurity, "research does not conclude that there is a strong positive relationship between warming and armed conflict" (IPCC 2014, 772). Detges (2017) also found mixed evidence, observing that studies tend to miss variables, use diverse indicators and conceptualizations, apply selective and limited evidence, and are biased toward statistical analyses. He argued this can lead to both over- and under-estimation of findings, misconceptions and flawed policies.

Hardt and Scheffran (2019) have observed that research is conflict- and Africa-focused, technocratic and managerial; lacks concepts of positive peace, Anthropocene/Earth Systems perspectives, as well as attention to power structures responsible for climate change; and has yet to prove clear correlations between climate and conflicts (7, 11–14). Mach et al. (2019) drew on expert interviews and literature to conclude that experts agree that climate does influence risks of intra-state violence, but its role is considered small compared to other drivers, and its mechanisms uncertain. Recently, the IPCC (2022) report has synthesized latest climate-security research, indicating, for example, that while climate impacts can drive insecurity, specific conflicts involve a wide range of factors.

Climate-security and the social sciences

In contrast to the study of variables and correlations, social scientific research has looked at the climate-security nexus as a social phenomenon. This work builds mainly on Buzan et al.'s work on securitization, which identified the environmental sector as one area of securitization, characterized by a strong scientific dimension and dramatic securitizations but limited success (1998, 74). They emphasized that, while climate change is a global phenomenon, its consequences tend to be local or regional, and this makes it hard to forecast the security complexes that might evolve around it (Ibid., 86). Building on Buzan et al., Jon Barnett (2001) considered that, while providing climate change with much-needed "gravitas", securitization has led to unproductive focus on militarization, sovereignty and threats from "unspecified others", instead of a more productive elevation of climate change from "low" to "high" politics, attention to human security and efforts to deal with global commons. Berling (2011) elaborated the role of science in securitization via a case study on climate security.

Further discourse-analytical studies include Trombetta (2008), which considered how security practices are being transformed by environmental issues, arguing that securitization of climate modifies those practices by challenging the realist tradition in IR, by creating demand for preventive and non-confrontational measures, and by enabling the participation of non-state actors in providing security. McDonald (2013) summarized how actors frame the relationship between climate change and security, including by comparing referent objects, agents of security, nature of threats, and responses. He considered which policy responses are more likely under different framings and concluded that the strongest framings are less likely to lead to effective responses to climate change. Comparing climate discourses, Diez et al. (2016) considered the similarities and differences between climate security discourses in Germany, Mexico, Turkey and the United States in 1990–2014.

Several studies have considered the drivers of climate-security discourses. For example, Brown et al. (2007) identified two drivers of climate securitization: worsening impacts, and efforts of European countries to mobilize support for a post-Kyoto climate deal using Darfur and Africa as cautionary tales. They criticized the tendency of climate-security discourses to apply anecdotal evidence and neo-Malthusianism, and highlighted challenges with separating climate from other factors and the poor predictive capacity of scarcity, urging caution about assuming a "straight-line progression" from scarcity to conflict (1148), and reminding that scaremongering can lead to climate fatigue and militarization, and distract from other challenges. Along similar lines, Verhoeven (2011) argued that efforts to link climate and security are driven by neo-Malthusianism, distract from political and economic dynamics in impacted countries, and can be instrumentalized by authoritarian regimes.

In addition, from an actor-network perspective, Mayer (2012) described how the framing of climate change has evolved from a long-term challenge to a national security nonlinear threat. He outlined the assemblage of practices, materials and actors that stabilized climate as such a threat and questioned the applicability of dichotomized approaches to study environmental governance and global politics. Mayer observed that climate-security tends to reinforce national interests, prioritize adaptation at the cost of mitigation and incentivize interventionism. Rothe (2017) studied how climate change is made manageable through assemblages of visual technologies (satellites, simulations and cooperation arrangements), which evolved from military hardware into broadly distributed tools, and how those assemblages can promote paternalistic control, favor the wealthy and silence discursive struggles.

These are the main contours of climate-security research, intended to illustrate the diversity of perspectives and findings. While the body of research is extensive, the evidence for climate-security links remains inconclusive and disputed, both in relation to correlations between climate parameters and conflicts, and with regard to whether specific conflicts such as Darfur, Syria or Lake Chad can be characterized as "climate wars". The IPCC, which has twice reviewed the evidence for climate-conflict links (2014, 2022), remains critical of such links. Social scientific studies have considered climate-security from a securitization and discourse-analytical perspective, examined its neo-Malthusian origins and mapped actor-networks relevant to climate-security links. Buzan et al. have observed that the securitization of environmental issues depends on whether policymakers follow the scientific agenda, the Paris Agreement reflects the global aspiration to follow the "best available science" when making climate policies, and climate activists have consistently urged the world to "follow the scientists". However, when it comes to climate-security, there is currently no scientific consensus, and the best available science, arguably captured in IPCC reports, has taken a critical view. Nevertheless, as the next sections will illustrate, efforts to securitize climate change keep intensifying.

Climate-security policy: acceleration regardless of controversies

Conversations about environmental security started in the 1970s and those about climate and security about a decade later, and intensified during the 1990s due to growing evidence for climate impacts and post-Cold War reframings of security (Buzan et al. 1998; Diez et al. 2016). This section presents some of the key milestones in climate-security policymaking at various forums and levels.

Climate and security as a national foreign and security policy issue

In the United States, articulations of climate as a security threat in the 1990s aimed to justify climate policies, such as the UNFCCC and its Kyoto Protocol and appeared in security strategies of George H. W. Bush and Clinton administrations (Diez et al. 2016, 42–4). They were eclipsed by G. W. Bush's climate skepticism and war on terror, but returned when a Pentagon study outlined dramatic scenarios for abrupt climate change (Schwartz and Randall 2003). Later, the Obama administration declared climate change a security policy priority, and the National Intelligence Council described how it might lead to Arctic confrontations, resources strains and increased migration (NIC 2008, xi). The 2010 National Security Strategy (The White House 2010) grouped climate change with nuclear proliferation, weapons of mass destruction, terrorism and piracy, describing how it leads to conflicts "over refugees and resources" (Ibid., 47). Later, President Obama asked government agencies to consider climate in national security measures (The White House 2016). The Trump administration sought to ignore the issue and proposed a panel to question assessments identifying climate as a national security risk (Gardner 2019). Recently, President Biden's 2021 Leaders' Summit on Climate featured a session on climate security (US Department of State 2021).

In the UK, Prime Minister Blair promoted the climate-security agenda in the news media (Blair 2004), and convened an expert symposium on "Avoiding Dangerous Climate Change" (Schellnhuber et al. 2006). In 2006, Secretary of Defense John Reid warned at Chatham House about the security threat of climate change (Russell and Morris 2006), Foreign Secretary Beckett identified "climate security" as an objective of UK foreign policy (Beckett 2006), and opposition leader David Cameron wrote about it in the *Financial Times* (Cameron 2007). The UK's first National Security Strategy (Cabinet Office 2008) described climate change as "potentially the greatest challenge" to security, and a similar strategy in 2015 listed it as a factor of instability (HM Government 2015).

In Germany, a 2007 report by the government advisory board WBGU stipulated that climate change "will draw ever-deeper lines of division and conflict in IR, triggering numerous conflicts between and within countries" (WBGU 2007, 1). Since then, Germany has pursued the climate-security agenda through its G7 presidencies, by promoting a UNSC role, by operating the Climate Diplomacy initiative and by organizing the Berlin Climate and Security Conference (BCSC).

Climate and security at the UN: debates and institutionalization

Focus on climate-security links has also increased at the UN. On 17 April 2007, the UNSC held its first debate on climate change (UNSC 2007(a), 2007(b)), and returned to the issue in 2011, 2018, 2019 and 2021. Participants highlighted Lake Chad, West Africa, Sahel, Somalia, small islands and areas affected by ISIS as high-risk areas, and proposed, for example, a special representative for climate and security, information arrangements, early warning mechanisms, integration of climate into mandates of UN missions, and a permanent UNSC seat for Small Island Developing States. However, Russia and China criticized that a UNSC focus on climate would undercut UN division of labor, force it to consider issues outside its expertise, distract from "real" conflict triggers and militarize climate change. India emphasized that the UNFCCC should be the forum for climate due to its universality and principles (UNSC 2018, 2019). In 2018, Germany and Palau launched the Ground of Friends of Climate and Security, a coalition of 56 countries interested in a more systematic consideration of the issue. At the 2019 GA debate, Canada identified climate change as a great security challenge, a "risk amplifier", and a cause of conflict and unrest, and asked the UNSC to recognize the need to act. Similarly, Germany highlighted that climate impacts and resource conflicts mean that future wars will be climate wars. In the 2020 debate, about dozen heads of state mentioned links between climate and security, highlighting, for example, that climate change is a risk multiplier and a threat to security, and urging the UN to strengthen its climate-security work.

These debates were accompanied by various activities of UN high-level officials. For example, in 2007, SG Ban Ki Moon wrote in the *Washington Post* that the Darfur conflict was partly due to climate change, and that other countries in Africa face similar risks (Ki Moon 2007). In 2009, Ki Moon also issued a report on climate and security for the GA (UN SG 2009).

In terms of institutions, in 2018, three UN organizations, the Department of Political and Peacebuilding Affairs (DPPA), the UN Environment Programme (UNEP) and the UN Development Programme (UNDP), established the Climate Security Mechanism (UNCSM) to address climate-related security risks. The UNCSM has published briefings, assessment guidelines, and checklists, and has set up a Community of Practice on Climate Security, an informal group with members from 23 UN entities.

Climate and security at other international forums

Outside the UN, the Munich Security Conference (MSC) has dedicated high-level panels to climate-security in 2014, 2016, 2017, 2018, 2019 and 2020. These have focused on climate-security "hotspots", impacts on territoriality and militaries, climate refugees, economic consequences, impacts on democracy, as well as the potential of the Paris Agreement to alleviate climate-security risks. In 2015, the G8/G7 published the report "A New Climate for Peace" (Rüttinger et al. 2015), which identified climate change as a threat multiplier and listed seven "compound climate-fragility risks". At the same time, the G7 Working Group on Climate and Fragility was established. The G8/G7 foreign ministers' communiques in 2013 and 2015–2016 all referred to security implications of climate change, (G8 2013; G7 2015, 2016), though such references to climate were nearly absent during the Trump administration in 2017–2020.

In 2019, the Environmental Peacebuilding Association convened the First International Conference on Environmental Peacebuilding and the German Federal Foreign Office (GFFO), adelphi, and the Potsdam Institute for Climate Impact Research (PIK) organized the now annual BCSC, the first conference dedicated to climate-security links. It called for consideration of climate-security risks in planning, enhanced capacities, integration of climate in peacebuilding, a stronger UNCSM and UNSC attention (GFFO et al. 2019; Adelphi et al. 2020(a), 2020(b)).

Civil society and climate-security links

This intensification of policymaking has been accompanied by work of NGOs and think tanks (Diez et al. 2016, 46). In 1994, the Woodrow Wilson Center launched an Environmental Change and Security Program (Ibid., 42). In 2007, the Center for Naval Analysis (CNA) Military Advisory Board, a group of retired generals, was the first to refer to climate as a "threat multiplier" (CNA 2007). Many other US think tanks launched similar work, and climate-security reports were published regularly by, for instance, the Center for Strategic and International Studies and Center for New American Security (Campbell et al. 2007, Rogers and Gulledge 2010), Council on Foreign Relations (Busby 2007), American Security Project (2012), Center for Climate and Energy Solutions (Huebert et al. 2012), Center for American Progress and the Heinrich Böll Stiftung (Werz and Conley 2012). The first NGO dedicated to climate-security was the Center for Climate and Security (CCS) set up in 2010 as an information hub. In Germany, civil society work since 2001 has been led by adelphi (Diez et al. 2016, 85), whose climate-security work has focused, in particular, on climate and fragility risks around Lake Chad (Nagarajan et al. 2018; Vivekananda et al. 2019), and a "handbook" on climate and security (Pohl et al. 2020).

Some efforts have evolved into coalitions of civil society and government. In 2015, a civil society consortium prepared the G7 climate fragility study (Rüttinger et al. 2015) and launched an online information hub on climate-security risks (Diez et al. 2016). The Climate Diplomacy initiative, which involves the GFFO and adelphi, has issued several climate-security reports, and coordinates a Climate Security Expert Network and an information website. Similarly, since 2015, the think tank Clingendael leads the Planetary Security Initiative for the Dutch Ministry of Foreign Affairs. In 2019, another coalition of government and non-governmental actors established the IMCCS, a consortium of military and security experts.

Climate-security and military planning

Parallel to policymaking, many armed forces have considered climate-security links, with focus on new threats, climate refugees, crisis response needs and protecting infrastructure (The Economist 2021). Since *Abrupt Climate Change* (2003), the US military has considered such risks. In 2007, the US Army War College conference on "The National Security Implications of Global Climate Change" discussed how the military can reduce emissions and assist adaptation (SSI 2007; Pumphrey 2008). James Mattis, Donald Trump's defense secretary, identified climate as a "driver of instability" (The Economist 2018) and Lloyd Austin, head of the US Department of Defense (DoD) in the Biden administration, wrote how it impacts global security, operations, plans and installations. Such positions are increasingly integrated into strategies. In 2016, the Pentagon ordered all mission plans to consider effects of climate change (US DoD 2016), and, in 2021, established a Climate Working Group (The Economist 2021). The UK climate strategy indicates that its armed forces are preparing for global temperature increase of 2 to 4 degrees Celsius, and the UK Ministry of Defense published its "Climate Change and Sustainability Strategic Approach" in March 2021 (Ibid.).

This section has illustrated how governments, international organizations, NGOs and national armed forces have advanced the broader climate-security debate and launched related policy responses. However, as seen in the previous section of this chapter, the evidence for climate-security and climate-conflict links remains inconclusive. While the world has pledged in the Paris Agreement to use the best available science to guide policymaking, climate-security seems to be moving forward despite the scientific controversies. As discussed in Chapter 1, these contradictory messages indicate that the scientific debate and policymaking are disconnected in some way. In order to understand this contradictory situation, this book aims to enhance understanding of the formation of the climate-security nexus by considering the work of actors to link climate and specific conflicts. However, before considering how that analysis is to be conducted, the next chapter will highlight how climate-security discourses form a part of a broader trend of nexus formation in international politics.

Literature

Adelphi and Postdam Institute for Climate Impact Research (PIK) 2020(a): *Summary – Berlin Climate and Security Conference, Part I.* At: https://berlin-climate-security-conference.de/sites/berlin-climate-security-conference.de/files/documents/summary_bcsc_2020_part_i.pdf (accessed 30 May 2021).

Adelphi GmbH and Postdam Institute for Climate Impact Research (PIK) 2020(b): *Summary – Berlin Climate and Security Conference, Part II*. At: https://berlin-climate-security-conference.de/sites/berlin-climate-security-conference.de/files/documents/bcsc_2020_part_ii_online_summary_0.pdf (accessed 30 May 2021).

American Security Project 2012: *Climate Security Report*. Washington, DC. At: www.americansecurityproject.org/climate-security-report/ (accessed 30 May 2021).

Barnett, J. 2001: *Climate Change and Security*. Tyndall Center for Climate Change Research, Working Paper 7, October 2001.

Beckett, M. 2006: *Foreign Policy and Climate Security*. Speech on 24 October, Berlin. At: https://webarchive.nationalarchives.gov.uk/20061102073121/www.fco.gov.uk/servlet/Front/TextOnly?pagename=OpenMarket/Xcelerate/ShowPage&c=Page&cid=10070293 91647&to=true&a=KArticle&aid=1161588023142 (accessed 28 September 2009).

Berling, T.V. 2011: *Science and securitization: Objectivation, the authority of the speaker and mobilization of scientific facts*. In: Security Dialogue, 42(4–5), 385–97.

Blair, T. 2004: *Speech given by the prime minister on the environment and the 'urgent issue' of climate change*. The Guardian. 15 September. At: www.theguardian.com/politics/2004/sep/15/greenpolitics.uk (accessed 27 August 2020).

Brown, I.A. 2010: *Assessing eco-scarcity as a cause of the outbreak of conflict in Darfur: A remote sensing approach*. In: International Journal of Remote Sensing, 31(10), 2513–20.

Brown, L.R. 1977: *Redefining National Security*. Worldwatch Institute, Washington, DC.

Brown, O.; Hammill, A. and McLeman, R. 2007: *Climate change as the 'new' security threat: implications for Africa*. In: International Affairs, 83(6), 1141–54.

Buhaug, H. 2010: *Climate not to blame for African civil wars*. In: PNAS, 107(38), 16477–82.

Buhaug, H. 2015: *Climate-conflict research: Some reflections on the way forward*. In: WIREs Climate Change, 26, 269–75.

Buhaug, H.; Nordkvelle, J.; Bernauer, T.; Böhmelt, T.; Brzoska, M.; Busby, J.W.; Ciccone, A.; Fjelde, H.; Gartzke, E.; Gleditsch, N.P.; Goldstone, J.A.; Hegre, H.; Holtermann, H.; Koubi, V.; Link, J.S.A.; Link, P.M.; Lujala, P.; O'Loughlin, J.; Raleigh, C.; Scheffran, J.; Schilling, J.; Smith, T.G.; Theisen, O.M.; Tol, R.S.J.; Urdal, H. and von Uexkull N. 2014: *One effect to rule them all? A comment on climate and conflict*. In: Climatic Change, 127, 391–7.

Burke, M.B.; Miguel, E.; Shanker, S., Dykema, J.A. and Lobell, D.B. 2009: *Warming increases the risk of civil war in Africa*. In: PNAS, 106(49), 20670–4.

Busby, J.W. 2007: *Climate Change and National Security – An Agenda for Action*. Council on Foreign Relations, CSR No. 32. Council of Foreign Relations Press, New York.

Buzan, B.; Waever, O. and de Wilde, J. 1998: *Security – A New Framework for Analysis*. Lynne Rienner Publishers Inc., London.

Cabinet Office 2008: *The National Security Strategy of the United Kingdom – Security in an Interdependent World*. Crown Copyright 2008. At: https://assets.publishing.service.gov.uk/government/uploads/system/uploads/attachment_data/file/228539/7291.pdf (accessed 20 December 2020).

Cameron, D. 2007: *A warmer world is ripe for conflict and danger*. Financial Times, 24 January, 15.

Campbell, K.M.; Gulledge, J.; McNeill, J.R.; Podesta, J.; Ogden, P.; Fuerth, L.; Woolsey, R. J.; Lennon, A.T.J.; Smith, J.; Weitz, R. and Mix, D. 2007: *The Age of Consequences: The Foreign Policy and National Security Implications of Global Climate Change*. Center for Strategic and International Studies and Center for New American Security. At: www.csis.org/analysis/age-consequences (accessed 28 September 2020).

Caruso, R.; Petrarca, I. and Ricciuti, R. 2014: *Climate Change, Rice Crops and Violence. Evidence from Indonesia.* CESifo Working Paper, No. 4665. Ifo Institute – Leibniz Institute for Economic Research at the University of Munich, Munich.

CNA Corporation 2007: *National Security and the Threat of Climate Change.* CNA Corporation, Washington, DC.

Corry, O. 2014: *The rise and fall of the global climate polity.* In: Stripple, J. and Bulkeley, H. (Eds.): Governing the Climate: New Approaches to Rationality, Power and Politics. Cambridge University Press, Cambridge, UK, 219–34.

Couttenier, M. and Soubeyran, R. 2014: *Drought and civil war in sub-Saharan Africa.* In: The Economic Journal, 124(575), 201–44.

Daoust, G. and Selby, J. 2022: *Understanding the politics of climate security policy discourse: The case of the Lake Chad Basin.* In: Geopolitics, 2, 1285–1322.

De Châtel, F. 2014: *The role of drought and climate change in the Syrian uprising: Untangling the triggers of the revolution.* In: Middle Eastern Studies, 50(4), 521–35.

Detges, A. 2016: *Local conditions of drought-related violence in sub-Saharan Africa: The role of road and water infrastructures.* In: Journal of Peace Research, 53(5), 696–710.

Detges, A. 2017: *Climate and Conflict: Reviewing the Statistical Evidence – a Summary for Policymakers.* Adelphi, Berlin.

Diez, T., von Lucke, F. and Wellmann, Z. 2016: *The Securitisation of Climate Change: Actors, Processes and Consequences.* Routledge Prio New Security Studies. Routledge, Oxon.

The Economist 2018: *Warriors and weather: Climate change and national security in America.* The Economist. 16 December. At: www.dailymotion.com/video/x6fbmah (accessed 25 November 2019).

The Economist 2021: *The men in green: The West's armies are getting more serious about climate change.* The Economist. 27 April. At: www.economist.com/international/2021/04/27/the-wests-armies-are-getting-more-serious-about-climate-change (accessed 12 August 2021).

Ehrlich, P.R. and Ehrlich, A.H. 1988: *The Environmental Dimensions of National Security.* Stanford Institute for Population and Resource Studies, Stanford, CA.

Falk, R. 1971: *This Endangered Planet: Prospects and Proposals for Human Survival.* Random House, New York.

Fjelde, H. and von Uexkull, N. 2012: *Climate triggers: Rainfall anomalies, vulnerability and communal conflict in Sub-Saharan Africa.* In: Political Geography, 31(7), 444–53.

Fröhlich, C. 2016: *Climate migrants as protestors? Dispelling misconceptions about global environmental change in pre-revolutionary Syria.* In: Contemporary Levant, 1(1), 38–50.

G7 Foreign Ministers 2015: *Communiqué of the G7 Foreign Ministers' Meeting.* Lübeck, Germany. 15 April. At: www.g7.utoronto.ca/foreign/150415-G7_Final_Communique.pdf (accessed 8 June 2021).

G7 Foreign Ministers 2016: *Communiqué of the G7 Foreign Ministers' Meeting.* Hiroshima, Japan. 10–11 April. At: www.g7.utoronto.ca/foreign/160411-communique.pdf (accessed 8 June 2021).

G7 Foreign Ministers 2021: *Foreign and Development Ministers' Meeting Communiqué.* London. 5 May. At: https://assets.publishing.service.gov.uk/government/uploads/system/uploads/attachment_data/file/983631/G7-foreign-and-development-ministers-meeting-communique-london-5-may-2021.pdf (accessed 23 June 2021).

G8 Foreign Ministers 2013: *G8 Foreign Ministers' Meeting Statement.* London. 10–11 April. At: www.g7.utoronto.ca/foreign/G8_Statement_Document_2013.pdf (accessed 8 June 2021).

Gardner, T. 2019: *White House readies panel to question security risks of climate*. Reuters. 20 February. At: www.reuters.com/article/us-usa-trump-climatechange/white-house-readies-panel-to-question-security-risks-of-climate-idUSKCN1Q92AR (accessed 9 June 2021).

Gerhardt, H.; Steinberg, P.E.; Tasch, J.; Fabiano, S.J. and Shields, R. 2010: *Contested sovereignty in a changing Arctic*. In: Annals of the Association of American Geographers, 100(4), 1–11.

GFFO, Adelphi, and PIK. 2019: *Berlin Call for Action*. 4 June. At: https://berlin-climate-security-conference.de/sites/berlin-climate-security-conference.de/files/documents/berlin_call_for_action_04_june_2019.pdf (accessed on 4 Jan 2020).

Gleick, P. 2014: *Water, drought, climate change, and conflict in Syria*. In: Weather, Climate and Society, 6(3), 331–40.

Hardt, J.N. and Scheffran, J. 2019: *Environmental Peacebuilding and Climate Change: Peace and Conflict Studies at the Edge of Transformation*. Toda Peace Institute Policy Brief No. 68. December 2019. At: https://toda.org/policy-briefs-and-resources/policy-briefs/environmental-peacebuilding-and-climate-change-peace-and-conflict-studies-at-the-edge-of-transformation.html (accessed 9 January 2020).

Her Majesty's (HM) Government 2015: *National Security Strategy and Strategic Defence and Security Review – A Secure and Prosperous United Kingdom*. Presented to Parliament by the Prime Minister by Command of Her Majesty. Crown Copyright, London.

Homer-Dixon, T. 1991: *On the threshold: Environmental changes as causes of acute conflict*. In: International Security, 16(2), 76–116.

Hsiang, S.M. and Burke, M.B. 2014: *Climate, conflict, and social stability: What does the evidence say?* In: Climatic Change, 123, 39–55.

Hsiang, S.M.; Burke, M.B. and Miguel, E. 2013: *Quantifying the influence of climate on human conflict*. In: Science, 341, 1235367.

Hsiang, S.M. and Meng, K.C. 2014: *Reconciling disagreement over climate-conflict results in Africa*. In: PNAS, 111(6), 2100–3.

Huebert, R.; Exner-Pirot, H.; Lajeunesse, A. and Gulledge, J. 2012: *Climate Change and International Security: The Arctic as a Bellwether*. Center for Climate and Energy Solutions, Arlington, VA. At: www.c2es.org/document/climate-change-international-security-the-arctic-as-a-bellwether/ (accessed 7 June 2021).

Ide, T. 2017: *Research methods for exploring the links between climate change and conflict*. In: Wiley Interdisciplinary Reviews: Climate Change, 8(3), e456.

IPCC 2014 (Adger, W.N.; Pulhin, J.M.; Barnett, J.; Dabelko, G.D.; Hovelsrud, G.K.; Levy, M.; Oswald Spring, Ú. and Vogel, C.H.): *Human security*. In: Field, C.B.; Barros, V.R.; Dokken, D.J.; Mach, K.J.; Mastrandrea, M.D.; Bilir, T.E.; Chatterjee, M.; Ebi, K.L.; Estrada, Y.O.; Genova, R.C.; Girma, B.; Kissel, E.S.; Levy, A.N.; MacCracken, S.; Mastrandrea, P.R. and White, L.L. (Eds.): Climate Change 2014: Impacts, Adaptation, and Vulnerability. Part A: Global and Sectoral Aspects. Contribution of Working Group II to the Fifth Assessment Report of the IPCC. Cambridge University Press, Cambridge and New York, NY, 755–91.

IPCC 2018 (Matthews, J.B.R. (Ed.)): *Annex I: Glossary*. In: Masson-Delmotte, V.; Zhai, P.; Pörtner, H.-O.; Roberts, D.; Skea, J.; Shukla, P.R.; Pirani, A.; Moufouma-Okia, W.; Péan, C.; Pidcock, R.; Connors, S.; Matthews, J.B.R.; Chen, Y.; Zhou, X.; Gomis, M.I.; Lonnoy, E.; Maycock, T.; Tignor, M. and Waterfield, T. (Eds.): Global Warming of 1.5°C: An IPCC Special Report on the Impacts of Global Warming of 1.5°C Above Pre-Industrial Levels and Related Global Greenhouse Gas Emission Pathways, in the Context of Strengthening the Global Response to the Threat of Climate Change, Sustainable Development, and Efforts to Eradicate Poverty. 539–62. At: www.ipcc.ch/site/assets/uploads/sites/2/2019/06/SR15_AnnexI_Glossary.pdf (accessed 5 July 2021).

IPCC 2022 (Pörtner, H.-O.; Roberts, D.C.; Tignor, M.; Poloczanska, E.S.; Mintenbeck, K.; Alegría, A.; Craig, M.; Langsdorf, S.; Löschke, S.; Möller, V.; Okem, A. and Rama, B. (Eds.): Climate Change 2022: Impacts, Adaptation, and Vulnerability. Contribution of Working Group II to the Sixth Assessment Report of the IPCC. Cambridge University Press, In Press. At: www.ipcc.ch/report/sixth-assessment-report-working-group-ii/ (accessed 19 June 2022).

Kelley, C.P.; Mohtadi, S.; Cane, M.A.; Seager, R. and Kushnir, Y. 2015: *Climate change in the fertile crescent and implications of the recent Syrian drought*. In: PNAS, 112(11), 3241–6.

Kevane, M. and Gray, L. 2008: *Darfur: Rainfall and conflict*. In: Environmental Research Letters, 3. At: https://iopscience.iop.org/article/10.1088/1748-9326/3/3/034006/pdf (accessed 21 November 2019).

Ki Moon, B. 2007: *A climate culprit in Darfur*. Washington Post. 16 June. At: www.washingtonpost.com/wp-dyn/content/article/2007/06/15/AR2007061501857.html (accessed 21 October 2019).

Linke, A.M.; O'Loughlin, J.; McCabe, T.J.; Tir, J. and Witmer, F.D.W. 2015: *Rainfall variability and violence in rural Kenya: Investigating the effects of drought and the role of local institutions with survey data*. In: Global Environmental Change, 34, 35–47.

Mach, K.J.; Kraan, C.M.; Adger, W.N.; Buhaug, H.; Burke, M.; Fearon, J.D.: Field, C.B.: Hendrix, C.S.; Maystadt, J.F.; O'Loughlin, J., Roessler, P.; Scheffran, J.; Schultz, K.A. and von Uexkull, N. 2019: *Climate as a risk factor for armed conflict*. In: Nature, 571, 193–7.

Maystadt, J.-F. and Ecker, O. 2014: *Extreme weather and civil war: Does drought fuel conflict in Somalia through livestock price shocks?* In: American Journal of Agricultural Economics, 96(4), 1157–82.

Mayer, M. 2012: *Chaotic climate change and security*. In: International Political Sociology, 6, 165–85.

McDonald, M. 2013: *Discourses of climate security*. In: Political Geography, 33, 42–51.

Myers, N. 1993: *Environmental refugees in a globally warmed world*. In: BioScience, 43(11), 752–61.

Myers, N. 1995: *Environmental Exodus: An Emergent Crisis in the Global Arena*. The Climate Institute, Washington, DC.

Nagarajan, C.; Pohl, B.; Rüttinger, L.; Sylvestre, F.; Vivekananda, J.; Wall, M. and Wolfmaier, S. 2018: *Climate-Fragility Profile: Lake Chad Basin*. Adelphi, Berlin. At: www.adelphi.de/en/system/files/mediathek/bilder/Lake%20Chad%20Climate-Fragility%20Profile%20-%20adelphi_0.pdf (accessed 21 November 2019).

National Intelligence Council (NIC) 2008: *Global Trends 2025: A Transformed World*. NIV, Washington, DC.

Okpara, U.T.; Stringer, L.C. and Dougill, A.J. 2016: *Perspectives on contextual vulnerability in discourses of climate conflict*. In: Earth System Dynamics, 7, 89–102.

Pohl, B.; Vivekananda, J.; Foong, A.; Wright, E.; Ivleva, D.; van Ackern, P.; Detges, A.; Rüttinger, L. and Smith, J.R. 2020: *Climate and Security – The Handbook*. Adelphi GmbH, Berlin. At: https://climate-diplomacy.org/sites/default/files/2020-10/Climate%20Security%20Handbook%20-%20adelphi_0.pdf (accessed 7 June 2021).

Pumphrey, C. (Ed.) 2008: *Global Climate Change: National Security Implications*. Strategic Studies Institute, U.S. Army War College, Carlisle, PA.

Rogers, W. and Gulledge, J. 2010: *Lost in Translation: Closing the Gap Between Climate Science and National Security Policy*. Center for New American Security, Washington, DC.

Rothe, D. 2017: *Seeing like a satellite: Remote sensing and the ontological politics of environmental security*. In: Security Dialogue, 48(4), 334–53.

Russell, B. and Morris, N. 2006: *Armed forces are put on standby to tackle threat of wars over water*. The Independent. 28 February. At: www.independent.co.uk/environment/armed-forces-are-put-on-standby-to-tackle-threat-of-wars-over-water-6108139.html (accessed 28 September 2020).

Rüttinger, L.; Smith, D.; Stand, G.; Tänzler, D. and Vivekananda, J. 2015: *A New Climate for Peace – Taking Action on Climate and Fragility Risks*. An Independent Report Commissioned by the G7 Members and Prepared by Adelphi, International Alert, Woodrow Wilson International Center for Scholars, and the EU Institute for Security Studies, Berlin.

Salehyan, I. and Hendrix, C. 2014: *Climate shocks and political violence*. In: Global Environmental Change, 28, 239–50.

Schellnhuber, H.J.; Cramer, W.; Nakicenovic, N.; Wigley, T. and Yohe, G. (Eds.) 2006: *Avoiding Dangerous Climate Change*. Cambridge University Press, Cambridge.

Scheffran, J.; Brzoska, M.; Kominek, J.; Lin, P.M. and Schilling, J. 2012: *Climate change and violent conflict*. In: Science, 336, 869–71.

Schleussner, C.-F.; Donges, J.F.; Donner, R.V. and Schellnhuber, H.J. 2016: *Armed-conflict risks enhanced by climate-related disasters in ethnically fractionalized countries*. In: PNAS, 113(33), 9216–21.

Schwartz, P. and Randall, D. 2003: *An Abrupt Climate Change Scenario and Its Implications for United States National Security*. Washington, DC. At: https://training.fema.gov/hiedu/docs/crr/catastrophe%20readiness%20and%20response%20-%20appendix%202%20-%20abrupt%20climate%20change.pdf (accessed 6 August 2021).

Selby, J.; Dahi, O.; Fröhlich, C. and Hulme, M. 2017(a): *Climate change and the Syrian civil war revisited*. In: Political Geography, 60(1), 232–44.

Selby, J.; Dahi, O.; Fröhlich, C. and Hulme, M. 2017(b): *Climate change and the Syrian civil war revisited: A rejoinder*. In: Political Geography, 60(1), 253–5.

Selby, J.; Daoust, G. and Hoffmann, C. 2022: *Divided Environments: An International Political Ecology of Climate Change, Water and Security*. Cambridge University Press, Cambridge.

Selby, J. and Hoffmann, C. 2014: *Beyond scarcity: Rethinking water, climate change and conflict in the Sudans*. In: Global Environmental Change, 29, 360–70.

Strategic Studies Institute of the U.S. Army War College (SSI) 2007: *The National Security Implications of Global Climate Change*. U.S. Army War College, Carlisle Barracks, PA. 30–31 March. At: https://web.archive.org/web/20070705034627/www.strategicstudiesin stitute.army.mil/events/details.cfm?q=82 (accessed 30 May 2021).

Theisen, O.M.; Gleditsch, N.P. and Buhaug, H. 2013: *Is climate change a driver of armed conflict?* In: Climatic Change, 117(3), 613–25.

Trombetta, M.J. 2008: *Environmental security and climate change: Analysing the discourse*. In: Cambridge Review of International Affairs, 21(4), 585–602.

Ullman, R.H. 1983: *Redefining security*. In: International Security, 8(1), 129–53.

UN 1992: *United Nations Framework Convention on Climate Change*. At: https://unfccc.int/resource/docs/convkp/conveng.pdf (accessed 5 July 2021).

United Nations Development Programme 1994: *Human Development Report*. Oxford University Press, New York. At: https://hdr.undp.org/system/files/documents/hdr1994 encompletenostatspdf.pdf (accessed 15 April 2022).

United Nations Secretary-General (UN SG) 2009: *Climate Change and Its Possible Security Implications*. Document A/64/350. United Nations. At: https://digitallibrary.un.org/record/667264.

United States Department of Defense (US DoD) 2016: *DoD Directive 4715.21 – Climate Change Adaptation and Resilience*. At: https://www.adaptationclearinghouse.org/

resources/u-s-department-of-defense-directive-4715-21-climate-change-adaptation-and-resilience.html.

United States Department of State 2021: *Leaders' Summit on Climate: Breakout Session on Climate Security*. 22 April. At: www.youtube.com/watch?v=6xa7yyypznY&t=22967s (accessed 8 June 2021).

UNSC 2007(a): *Security Council Holds First-Ever Debate on Impact of Climate Change on Peace, Security, Hearing Over 50 Speakers*. UNSC Press Release, SC/9000. 17 April. At: www.un.org/press/en/2007/sc9000.doc.htm (accessed 20 August 2020).

UNSC 2007(b): *Letter Dated 5 April 2007 from the Permanent Representative of the United Kingdom of Great Britain and Northern Ireland to the UN Addressed to the President of the Security Council*. S/2007/186. 5 April. At: www.securitycouncilreport.org/atf/cf/%7B65BFCF9B-6D27-4E9C-8CD3-CF6E4FF96FF9%7D/Ener%20S%202007%20186.pdf (accessed 29 September 2020).

UNSC 2011(a): *Record of the 6587th Meeting*. Document S/PV.6587. At: https://undocs.org/en/S/PV.6587 and https://undocs.org/en/S/PV.6587(Resumption1) (accessed 20 December 2020).

UNSC 2011(b): *Statement by the President of the Security Council*. Document S/PRST/2011/15. At: https://undocs.org/en/S/PRST/2011/15 (accessed 20 December 2020).

UNSC 2018: *Addressing Security Council, Pacific Island President Calls Climate Change Defining Issue of Next Century, Calls for Special Representative on Issue*. UN Press Release SC/13417 on the 8307th Meeting of the Security Council. At: www.un.org/press/en/2018/sc13417.doc.htm (accessed 30 May 2015).

UNSC 2019: *Massive Displacement, Greater Competition for Scarce Resources Cited as Major Risks in Security Council Debate on Climate-Related Threats*. UN Meetings Coverage. 8451st meeting of the Security Council. 25 January. At: www.un.org/press/en/2019/sc13677.doc.htm (accessed 7 January 2020).

UNSC 2021: *Maintenance of International Peace and Security: Climate and Security – Security Council Open VTC*. 23 February. At: http://webtv.un.org/search/maintenance-of-international-peace-and-security-climate-and-security-security-council-open-vtc/6234686966001/?term=&lan=english&cat=Security%20Council&sort=date&page=5 (accessed 29 May 2021).

Verhoeven, H. 2011: *Climate change, conflict and development in Sudan: Global neo-Malthusian narratives and local power struggles*. In: Development and Change, 42(3), 679–707.

Vivekananda, J.; Wall, M.; Sylvestre, F. and Nagarajan, C. 2019: *Shoring Up Stability: Addressing Climate and Fragility Risks in the Lake Chad Region*. Adelphi, Berlin. At: www.adelphi.de/en/publication/shoring-stability (accessed 23 November 2019).

Von Uexkull, N. 2014: *Sustained drought, vulnerability and civil conflict in Sub-Saharan Africa*. In: Political Geography, 43, 16–26.

Von Uexkull, N. 2016: *Climate, Conflict and Coping Capacity: The Impact of Climate Variability on Organized Violence*. Doctoral thesis, Uppsala University, Disciplinary Domain of Humanities and Social Sciences, Faculty of Social Sciences, Department of Peace and Conflict Research. At: www.diva-portal.org/smash/record.jsf?pid=diva2%3A951030&dswid=1989 (accessed 4 September 2021).

Werrell, C.E.; Femia, F. and Sternberg, T. 2015: *Did we see it coming?: State fragility, climate vulnerability, and the uprisings in Syria and Egypt*. In: SAIS Review of International Affairs, 35(1), 29–46.

Werz, M. and Conley, L. 2012: *Climate Change, Migration, and Conflict. Addressing Complex Crisis Scenarios in the 21st Century*. Center for American Progress and Heinrich Böll Stiftung. At: www.americanprogress.org/wp-content/uploads/issues/2012/01/pdf/climate_migration.pdf (accessed 30 May 2021).

The White House 2010: *National Security Strategy*. Washington, DC. May. At: https://oba
mawhitehouse.archives.gov/sites/default/files/rss_viewer/national_security_strategy.pdf
(accessed 3 June 2021).

The White House 2016: *Presidential Memorandum – Climate Change and National Security*. At: https://obamawhitehouse.archives.gov/the-press-office/2016/09/21/presidential-
memorandum-climate-change-and-national-security (accessed 25 November 2019).

Wissenschaftlicher Beirat der Bundesregierung Globale Umweltveränderungen (WBGU)
2007: *Welt im Wandel: Sicherheitsrisiko Klimawandel*. Springer, Berlin.

Zhang, D.D.; Brecke, P.; Lee, H.F.; He, Y. and Zhang, J. 2007: *Global climate change, war,
and population decline in recent human history*. In: PNAS, 104, 19214–19.

3 Nexus formation, knowledge and international relations

Issue-linkages and nexuses in international relations

Chapter 2 outlined the connections between climate and security in science and policymaking. It highlighted the contradictory co-existence of commitments to the best available science, disputed evidence about climate-security links and accelerating securitization. It also suggested that deep empirical analysis of the role of knowledge in climate-security debates can help understand the possible disconnect between science and policymaking.

Before considering how such analysis can be conducted, this chapter contextualizes climate-security debates by elaborating on how they form a part of a broader trend of nexus formation in international politics. The chapter begins by outlining the patterns of nexus formation in contemporary governance. This is followed by a discussion of how existing research, including in IR, has approached the phenomena of formation of issue-linkages and nexuses in international politics, and elaborates how this study will contribute to that research landscape by describing the processes to merge previously disconnected issues – in this case climate and security.

From silos to nexuses

In 2013, Bruno Latour observed what happens to an anthropologist studying a "modern" society:

> The Moderns present themselves to her in the form of domains, interrelated, to be sure, but nevertheless distinct: Law, Science, Politics, Religion, The Economy, and so on; and these, she is told, must by no means be confused with one another . . . When one is "in Science", she is assured, one is not "in Politics", and when one is "in Politics", one is not "in Law", and so forth.
>
> (Latour 2013, 29)

National and international governance has strongly reflected such compartmentalization, with international organizations working on specific domains such as telecommunications (International Telecommunications Union), postal services

DOI: 10.4324/9781003451525-3

(Universal Postal Union), aviation (International Civil Aviation Organization), food and agriculture (FAO), health (World Health Organization), atomic energy (International Atomic Energy Agency), the environment (UNEP), climate change (UNFCCC), biodiversity (Convention on Biological Diversity) and security (UNSC). And the phenomenon is not limited to UN organizations. Dozens of sectoral, regional, national subnational and local organizations steer policy and research within domains such as agriculture, fisheries, maritime issues, finance, trade, customs, culture, transportation, law enforcement, arms control and defense.

However, such compartments are not cast in stone. Linking and merging of issues is a constant feature of global politics (Corry 2014; Stripple and Bulkeley 2014; Allan 2017), and, inspired by earth systems sciences, there has been increased attention to interconnected systems instead of fixed domains. James Lovelock's Gaia thesis considered the Earth to be a single self-regulating organism (see e.g. Lovelock 2000). In 2001, the Amsterdam Declaration, issued by a conference of four research coalitions, described the Earth as a single, self-regulating system with physical, chemical, biological and human components involving complex interactions and variabilities. In 2005, the *Journal of Earth Systems Science* was established to capture research of such interconnected nature. In 2009, scientists identified nine planetary boundaries, an interconnected set of limits of the earth system within which humans can safely operate (Corry 2014).

Such observations in the scientific realm have inspired a shift toward "nexus approaches" in international governance. In 2018, Liu et al. argued, in the context of various global challenges, that when interconnected problems are addressed individually, dealing with one problem can make others worse, and that "silo" approaches used by specialized organizations "cannot effectively address the linked challenges" (2018, 466). The authors called for "broad, multi-sector, multi-scale and multi-regional perspective to solve global challenges", awareness of both harmful and positive "trade-offs", and "synergies when solving major problems" (Ibid.). Nexus approaches were presented as solutions that help identify synergies, detect harmful trade-offs, reveal unexpected consequences, promote integrated governance, enhance cooperation, reduce conflicts, increase efficiency and reduce waste and pollutants (Ibid., 474). To strengthen nexus governance, Liu et al. called for the consideration of more sectors, scales and places; integration of overlooked drivers and regions; the development of "nexus toolboxes", and making nexus strategies central to governance for "integrated SDG implementation" (Ibid., 466).

As part of this move toward nexus approaches, organizations and individuals developed research, publications and institutions to establish and govern multiple nexuses, including climate-security, climate-migration, climate-health, climate-poverty, climate-water, climate-energy, climate-biodiversity or the climate-land-water-energy-development nexus (see e.g. UN-DESA 2014; Carling 2017; Zelli et al. 2020), to mention a few.

But what exactly does it mean to build a nexus between two pre-existing areas of governance? How exactly should we understand and operationalize such merging processes? And what does IR research say about them? The next section synthesizes how existing IR research has considered nexus formation and related phenomena.

Issue linkages and nexuses in existing research

The first insights to the formation of linkages and nexuses between governance areas in IR were articulated by Ernst Haas, who, when considering interdependence and regime formation, paid attention to how previously separate issues are linked into packages he called "issue areas", and how issue areas might become the basis for regime formation. Haas observed that when governments realize complex linkages between previously separate issues, those issues might evolve into issue areas (1980, 361).

As examples, Haas described how an international "monetary issue-area" emerged in 1944 from previously separate considerations of exchange rates, economic growth and inflation after participants realized that each had impacts on the others, and that all influenced the overall economic welfare of the involved countries (Ibid., 364–5). He also outlined how the oceans, previously considered under separate domains of fisheries, transport, resource extraction and research, became an issue area in 1967 when ocean questions were linked with sovereign equality of states, and interdependencies among the various ocean activities were recognized (Ibid., 366). But Haas also emphasizes that the cognitive recognition of issue-linkages does not automatically lead to regime construction, and that linkages might also decline over time. He describes how this happened with the UNCTAD negotiation that started as a large package in 1974, and was then disaggregated into many specific items (Ibid., 395).

In terms of the formation of issue areas, Haas considered that issues might be linked into issue areas in three different ways: first, negotiating tactics (to gain leverage); second, to maintain existing coalitions; and third, and most interestingly for this study, participants can also form what Haas called "substantive linkages" "on the basis of cognitive developments based on consensual knowledge linked to an agreed social goal" (Ibid., 372). The notion of substantive linkages is of particular relevance to the study of the formation of the climate-security nexus through knowledge resources because it recognizes the role of knowledge in the formation of issue areas.

While the study of issue-linkages began decades ago, the recent emerging of nexuses in development policy was described in an IR context in *Security Dialogue* by Stern and Öjendal. Taking the "development-security nexus" as a case study, the authors defined a nexus as a "network of connections between disparate ideas, processes or objects" (2010, 11). They observed that nexuses permeate contemporary development discourse, resonate in the minds of many actors, and have begun to attract resources and trigger institutional changes. However, Stern and Öjendal highlighted that there are multiple conceptual unclarities about nexuses, that critical considerations of the evocation of nexuses are often lacking, that the discourses tend to imply agreement about the content and consequences of nexuses without clarity, and that there has been limited academic attention to nexus formation.

In that spirit, Stern and Öjendal critically examined the concept of nexuses, mapped different articulations of the development-security nexus and considered the implications of those articulations on policies. They identified six storylines about the nexus, including that it: (1) is a modern teleological narrative that reflects expectations of progress of societies toward development and security; (2) broadens and humanizes development and security away from exploitative capitalist modes by putting humans in the center; (3) leads to an impasse, where both development and security remain perpetually out of reach; (4) reflects post-development mind-sets oriented against colonial attitudes inherent to development and security discourses; (5) is a technique of governmentality/biopower used to discipline poor countries; (6) reflects expectations of globalized development based on dealing with "interrelated and mutually constitutive human global survival issues" (Ibid., 22). Based on the mapping, the authors concluded that content and form of the development-security nexus are not clear, nexuses mean wildly different things to different actors, and they are thus open to all kinds of uses, including problematic ones, under the guise of progressive politics (e.g. descriptions of poverty as a disease that causes conflict can justify oppressive control to prevent the spreading of such problems to "healthy" countries).

In addition to the analysis of issue-linkages and nexuses within IR frameworks, several publications from development and sectoral governance perspectives have considered emerging nexus dynamics. Nexus thinking, which is particularly pertinent in the development sector, assumes that due to interlinkages between areas of governance, those areas need to be considered in an integrated fashion (see Carling 2017; Liu et al. 2018; Zelli et al. 2020). Many international conversations are happening to articulate various kinds of nexuses, developing methods to understand them and establishing processes to consider issues from a nexus perspective.

From a critical perspective, Cairns and Krzywoszynska consider nexus thinking in natural resource debates in the UK, describing "nexus" as a buzzword that lacks agreed definitions but nevertheless evokes strong normative resonance and can thus be appropriated to suit multiple agendas. They characterize nexus discourses as part of a global science-policy trend that fetishizes integration, focuses on technical solutions and efficiency gains and paves the way for "technocratic forms of environmental managerialism" (2016, 3).

Williams et al. (2018) explored the emerging water-energy nexus, and the governance models and implications inherent to nexus thinking. They described how nexuses emerged from environmental governance circles in response to sectoral competition and are now a key feature of development discourses (654). They considered nexus thinking as "technocratic and reductionistic" which tends toward neoliberal techno-managerial solutions focused on integration of nexus components and criticize the way the "panacea of integration" forestalls policy debates and prevents policy change (Ibid., 653–4). While nexus governance solutions usually aim at enhanced management, reduced negative trade-offs and amplified synergies, the authors observed that most contributions do not specify what such integration would actually look like. They considered that "the idea of integration has become a catholicon for the negative aspects of nexus integration, unquestioned and never problematized, but one that is consistently ill-defined" (Ibid., 660). The authors conclude that nexus thinking and its ideology of integration presents a paradox: they outline huge interconnected problems, accompanied at the same time by ready-made solutions which seem to require no real change (Ibid., 661). Nexus discourses preclude critical interpretations, turn policy challenges into technical questions and have thus evolved into tools for the depoliticization of nature (Ibid., 662).

These are some of the illustrations of how existing research has considered the formation of issue-linkages and nexuses, including from the perspective of IR and development research. To complement this landscape of nexus research, this book will consider a key part of the efforts to connect two previously disconnected hegemonial governance areas of climate and security together into what is often referred to as a climate-security nexus – the connecting of climate change and specific conflicts.

The role of knowledge in nexus formation

However, to understand the climate-security nexus, it is necessary to consider the production and use of knowledge. On the one hand, past IR research has drawn attention to the key role of knowledge and expertise in international politics (Haas 1980; Bueger 2014). On the other, climate-security links are particularly dependent on knowledge due to the way we know about climate change.

Knowledge: basis of climate policy

As stipulated by the IPCC, climate change is identifiable by observing changes in the *mean properties* of the climate, derived from *statistical tests* (see Chapter 2). Knowledge of the climate is created by a "vast machine" (Edwards 2013) of monitoring equipment, communication technologies, databases, statistical tools, scientific publications, IPCC reports and other such arrangements. Without these, humanity would neither know about climate change, nor have any basis for linking it with other governance areas. In contrast, conventional security threats tend to involve concrete, immediate and visible processes such as the military build-ups, terrorist attacks, or airspace violations. Those are generally perceived through intelligence operations, reconnaissance, satellites and other components of security apparatuses, rather than accumulation of statistical knowledge of mean parameters over extended periods of time. The increasing merging of issue areas such as climate and security means changing roles for knowledge and expertise. Rychnovská et al. observe that "with the rise of risk management practices . . ., natural scientists and other experts have become increasingly involved in security politics" (Rychnovská et al. 2017, 327).

Given this centrality of knowledge for climate-related issues, but also for the general movement toward risk management practices, this book will focus on the role of knowledge resources in linking climate and security. Through this, it will contribute to existing IR research by describing the role of such resources in the processes to merge the previously disconnected hegemonial governance areas of climate and security into a nexus.

IR research on knowledge, issue-linkages and nexus formation

But how has the role of knowledge resources in the formation of nexuses been approached by IR research? In the context of his consideration of the formation of issue-linkages in IR, Ernst Haas has also devoted attention to the role of knowledge. He has argued that the fragmented and fluctuating nature of regimes and issues means that evolving knowledge is a key factor in international politics:

> There are no structures, just aggregates linked by a changing . . . cause-effect chains . . . no global system, just sub-systems which tend to rearrange them-selves without central guidance . . . no overall complexity, merely successive and fallible human efforts to understand interdependence . . . no overriding evolutionary dynamic, only isolated and lonely thrusts into more elaborate forms of survival in one area of concern or another. The very ephemeral and temporary quality of these wholes, and of the fluctuating organizations to which such conceptions must give rise, depends heavily on the changing character of knowledge.
>
> (Haas 1975, 870–1)

More specific insights into the study of knowledge in forming issue-linkages can be found in Haas' 1980 work. Haas considered that knowledge is the "basic ingredient for exploring the development of issue-areas" (1980, 367), that consensual knowledge relevant to policy goals can lead to packages of issue-areas (Ibid., 372), and that the formation of issue-areas is usually preceded by consensus related to intellectual strategies or novel causal understandings (Ibid., 374).

As described earlier, while Haas acknowledged that issue-areas might be estab-lished as a negotiating tactic or to maintain existing coalitions, he also emphasizes that knowledge plays a limited role in such situations. However, he recognized the key role of knowledge in the formation of "substantive linkages", which means issue-areas that emerge when negotiators link issues based on emerging consensus on "some intellectual strategy of evolving causal understanding" (Ibid.). In other words, previously separate issues can be merged into issue areas when consensus knowledge develops about the linkages and interdependencies between them.

In addition, Haas also highlighted that the construction of substantive linkages always involves some ordering, and observed that, while sometimes "a persuasive doctrine lights up the murky world of uncertainty, and consensus develops around it – thus ordering the steps the rulers must take", ambiguity and uncertainty tend to be the norm, forcing actors to deal with the situation by "filtering, simplifying, and foreclosing choices", by "fudging" rather than systematically examining trade-offs, and by applying questionable past precedents to the situation at hand. Thus, order-ing tends to be chaotic and sporadic, and "causal sequences are assumed or guessed at rather than studied fully" when linking issues (Ibid., 378).

Haas also warned against technocratic determinism, and emphasized that knowl-edge is not a sufficient condition for the formation of issue-areas because existing considerations of national interest will not change only because new insights about certain issues are available (Ibid.). While experts might agree about causes, effects and linkages, this does not guarantee that policymakers will be impressed, as they are always driven by multiple concerns. Thus, Haas concludes:

> Substantive knowledge *alone* cannot legitimate a holistic package of issues. The legitimation depends on the acceptance of a new understanding on the part of key political actors. Governments – even when exposed to novel insights about energy, growth, pollution, or food – cannot be expected to stop considering their policies within the perspective of what passes for the national interest. Substantive issue-linkage depends on learning that the national interest can be redefined or broadened, and that international collaboration is required for the realization of national goals. Knowledge can legitimate collaborative behavior only when the possibility of joint gains from the collaboration exists and is recognized.
>
> (Ibid., 374)

> Learning to link issues substantively requires more than new scientific and technical information, and more than a formula that declares international equity to be the norm of order. It also requires a demonstration that only by linking issues substantively can everyone's goals be realized. If one is not an ecological holist, it may be impossible to sustain such a demonstration.
>
> (Ibid., 394)

This way, Haas identified knowledge as the "basic ingredient" of issue-linkages as well as international politics and regimes. This book aims to further understanding the role of knowledge resources, including scientific knowledge, in formation of issue-linkages and nexuses by highlighting the efforts through which actors try to achieve consensus about policy questions around climate-conflict links, or, in other words, the exact content and enabling resources of the "ephemeral and temporary wholes" described by Haas. While Haas looked at the impact of knowledge, this book is about the resources and circulation modalities that make knowledge about climate-conflict links possible in the first place. Understanding that dimension will require building on research approaches that are sensitive to the role of knowledge and to processes of linking disconnected things. The next chapter will describe the research strategy toward that end.

Literature

Allan, B.B. 2017: *From subjects to objects: Knowledge in International Relations theory.* In: European Journal of International Relations, 24(4), 841–64.

Bueger, C. 2014: *From expert communities to epistemic arrangements: Situating expertise in international relations.* In: Mayer, M.; Carpes, M. and Knoblich, R. (Eds.): International Relations and the Global Politics of Science and Technology. Springer VS, Wiesbaden, 39–54.

Cairns, R. and Krzywoszynska, A.D. 2016: *Anatomy of a buzzword: The emergence of 'the water-energy-food nexus' in UK natural resource debates.* In: Environmental Science and Policy, 64, 164–70.

Carling, J. 2017: *Thirty-Six Migration Nexuses, and Counting*. At: https://jorgencarling. org/2017/07/31/thirty-six-migration-nexuses-and-counting/ (accessed 9 September 2020).

Corry, O. 2014: *The rise and fall of the global climate polity*. In: Stripple, J. and Bulkeley, H. (Eds.): Governing the Climate: New Approaches to Rationality, Power and Politics. Cambridge University Press, Cambridge, 219–34.

Edwards, P.N. 2013: *A Vast Machine – Computer Models, Climate Data, and the Politics of Global Warming*. MIT Press, Cambridge, MA.

Haas, E.B. 1975: *Is there a hole in the whole? Knowledge, technology, interdependence, and the construction of international regimes*. In: International Organization, 29(3), 827–26.

Haas, E.B. 1980: *Why collaborate?: Issue-linkage and international regimes*. In: World Politics, 32(3), 357–405.

Latour, B. 2013: *An Inquiry into Modes of Existence – An Anthropology of the Moderns*. Harvard University Press, Cambridge, MA.

Liu, J.; Hull, V.; Godfray, C.; Tilman, D.; Gleick, P.; Hoff, H.; Pahl-Wostl, C.; Xu, Z.; Sun, J. and Li, S. 2018: *Nexus approaches to global sustainable development*. In: Nature Sustainability, 1, 466–76.

Lovelock, J. 2000: *Gaia: A New Look at Life on Earth*. Oxford University Press, Oxford.

Rychnovská, D.; Pasgaard, M. and Berling, T.V. 2017: *Science and security expertise: Authority, knowledge, subjectivity*. In: Geoforum, 84, 327–31.

Stern, M. and Öjendal, J. 2010: *Mapping the security development nexus: Conflict, complexity, cacophony, convergence?* In: Security Dialogue, 41(1), 5–30.

Stripple, J. and Bulkeley, H. 2014: *On governmentality and climate change*. In: Stripple, Johannes and Bulkeley, H. (eds.): Governing the Climate: New Approaches to Rationality, Power and Politics. Cambridge University Press, Cambridge, UK, 1–24.

United Nations Department of Economic and Social Affairs (UN-DESA) 2014: *Prototype global sustainable development report*. New York. At: https://sustainabledevelopment. un.org/index.php?page=view&type=400&nr=1454&menu=35 (accessed 3 October 2022).

Williams, J.; Bouzarovski, S. and Swyngedouw, E. 2018: *The urban resource nexus: On the politics of relationality, water – energy infrastructure and the fallacy of integration*. In: Politics and Space, 37(4), 652–69.

Zelli, F.; Bäckstrand, K.; Nasiritousi, N.; Skovgaard, J. and Widerberg, O. 2020: *Governing the Climate-Energy Nexus: Institutional Complexity and Its Challenges to Effectiveness and Legitimacy*. Cambridge University Press, Cambridge.

4 Understanding knowledge and climate-conflict links with a sociology of translations

Understanding knowledge and the climate-security nexus through a sociology of translations

Chapter 2 described climate-security research and policymaking, illustrated how the combination of disputed evidence and accelerating securitization indicates a disconnect between knowledge and policymaking, and highlighted the need for empirical studies into the role of knowledge. Chapter 3 contextualized the linking of climate and security as part of a trend of nexus formation in IR, and emphasized that the climate-security nexus can be understood with a research strategy that is sensitive to the role of knowledge and processes of linking things. To that end, this study draws insights from ANT to articulate a research strategy referred to as a *sociology of translations*. ANT studies were developed to study the linking of domains through translations, and the role of knowledge in social processes. This chapter presents ANT, how it has been used in IR research, and the rationale for adopting it. It outlines the guiding concepts and principles of the analysis, the empirical basis, operationalization, and limitations.

Actor-network theory

ANT is a branch of science and technology studies (STS), a set of research approaches focused on the role of science and technology in society. STS considers the boundary between science and policy blurred, complex and open to many framings (Irwin 2008; Bijker et al. 2009; Boswell 2009), questioning the common sense perspective that science and policy are separate domains (Weingart 1999; Jasanoff 2004; Latour 2005; Boswell 2009). STS focus on complexities of production of knowledge, assuming that facts are constructed through processes involving theories, experiments, data, debates, publication and many other factors (Latour 1987, 2004, 2005; Jasanoff 2004; Erickson 2016). Thus, STS is in particular a critique of traditional science-society dichotomies.

STS is methodologically dynamic and diverse, drawing on history, philosophy, sociology, political science, law, economics, and anthropology (Jasanoff 2004). It studies phenomena by "following the actors" while avoiding a priori judgements about issues and entities (Latour 2005; Irwin 2008). STS scholars study structures,

DOI: 10.4324/9781003451525-4

practices, ideas, materials and linkages between knowledge, culture and power. Common focus areas include the co-production of knowledge, how the boundary between science and other domains is negotiated, as well as networks and assemblages (Ibid.).

Co-productionist approaches assume that science and policy develop together: "Knowledge-making is incorporated into practices of state-making, . . . and . . . practices of governance influence the making and use of knowledge" (Jasanoff 2004). The interest is in how knowledge and political action "have become mutually embedded and co-constituted" (Irwin 2008, 586). Boundary approaches study activities that demarcate science from other domains. Although science and policy cannot be separated by objective criteria, boundary work defines things as scientific or non-scientific (Ibid.). The third perspective – networks and assemblages – focuses on how "human and nonhuman actors are enrolled in the construction of sociotechnical systems" (Ibid., 592). From this perspective, "governance" involves diverse organizing mechanisms, operational assumptions, modes of thought, and practical activities. Power does not reside only in institutions or powerful individuals, but in "de-centered networks and shifting assemblages of power" (Ibid., 584).

The research strategy of this book draws on the study of networks, and specifically ANT. ANT was developed by Bruno Latour, John Law, Michel Callon and others.[1] Initially, ANT studies focused on scientific practice and science-society interactions, and, more specifically, for example, on scallop rehabilitation, bush pumps and atherosclerosis. Law defined ANT as a "family of material-semiotic tools, sensibilities and methods . . . that treat everything in the social and natural worlds as a continuously generated effect of the webs of relations within which they are located" (Law 2009, 141). This means that everything that has social existence is an enactment of a web of relations. This is a result of the principle of relationality – the basic ontological position of ANT: actors acquire identities and roles only via associations with others (Braun et al. 2019). ANT studies those webs of relations and their effects.

The webs of relations studied through ANT are heterogeneous. They can include humans, technical arrangements, natural phenomena, methodologies, materials, animals and so forth – anything that makes a difference in actions or the structure of the web. The webs can be fragile and change. ANT's challenge is to describe the heterogeneous webs in all their fragility and obduracy. ANT is particularly interested in (based on Callon 1986; Latour 2005; Fuller 2006; Law 2009; Mol 2010):

- How, and out of which bits and pieces, relations between actors are assembled, and how those relations do or do not hold together;
- How the enactment of new associations produces and reshuffles various kinds of actors;
- The productive role of materials, technologies, and practices in the webs of relations;
- How the webs expand, shrink or are displaced due to circulation of information;
- How the "mode of ordering" of a web of relations compels actors to change behavior and enables others to steer behavior – in other words, how a web generates power and influence.

In addition, ANT is often described as a tool to "sensitize" researchers to help uncover unexpected aspects of an issue (Mol 2010), to avoid adding unnecessary context (Fuller 2006), and to reduce the risk of letting simplified social theories drive the analysis (Law 2009).

ANT has many origins, including, for example, De Saussure's semiotics of language (Mol 2010, 257), Serres' concept of translation (Law 2009) and Foucault's approach to social order (Law 2009; Mol 2010). But it has also evolved through empirical studies – often referred to as "classics" – which constitute a "toolbox" for ANT. The classics all consider different phenomena, but share the focus on relationality, webs of human and nonhuman actors, and materials and practices (Bueger and Stockbruegger 2017).

In IR research, ANT has inspired studies of environmental security, security networks, torture and practices within global economic networks. As described in Chapter 2, Mayer (2012) considered climate-security links from an ANT perspective, outlining the assemblage that stabilized climate change as a security threat, and the trends of reinforcement of national interests, prioritization of adaptation and incentives for interventionism. Rothe (2017) studied how climate is made manageable with satellites and simulations, and outlined risks of paternalistic control, favoring the rich and silencing discursive struggles. Aradau (2010) considered the securitization of critical infrastructure as an interaction of material and discursive practices. On torture, Austin (2015) described the Argentinian "Death Flights" and the US extraordinary rendition program and argued that material 'allies' such as airplanes, sedatives and military bases enable torture and circumvented conventional democratic-autocratic binaries. Austin (2016) also elaborated on how torture is enabled by global technological networks. Callon (2007) focused on the performativity of economics and the development of the *homo oeconomicus* through practices of economics. Henriksen (2013) studied how the World Bank constructed a global market for microfinance through performance indicators, standardizations and calculation devices of neo-classical economics.

Researchers have observed that ANT can enrich IR and security studies by:

- Drawing attention to dynamic processes, relations and practices at work in international politics, in contrast to static conditions (Nexon and Pouliot 2013, 342);
- Finding solutions to the agent-structure problem by approaching actors and structures as mutually constitutive and dynamic (Ibid., Braun et al. 2019);
- Differentiated considerations of what it means to act, beyond the "usual suspects" (states, organizations and leaders) (Braun et al. 2019);
- Documenting how power is generated via networks of human and nonhuman actors and the structures of their associations (Best and Walters 2013(a), 333), and how actors considered powerful depend on "non-human instruments and inscription devices" (Barry 2013, 415);
- Studying expert authority in international contexts (Best and Walters 2013 (b), 347).

A sociology of translation to study knowledge in nexus formation

The sociology of translations adopted for this study is not based on a fixed template, but rather on precedents, concepts and principles drawn from ANT studies. The principle of relationality means that climate-security links are an effect of webs of relations established by actors. This book describes how those relations are integrated within the climate-security debates into an actor-network, the roles of materials and practices (in particular knowledge resources) in the actor-network, and how the modes of ordering of the network influence behavior.

Why a sociology of translations?

ANT-based approaches help study the role of knowledge resources in the formation of the climate-security nexus for the following reasons:

- They have been developed for and applied to considering how knowledge is assembled for social processes in their complexity and heterogeneity. Thus, they enable a differentiated consideration of knowledge. This is particularly relevant in light of the hybrid nature of the "vast machine" of climate sciences (Edwards 2013) that feeds climate-security discourses.
- The concept of translation enables looking how previously separate things are associated together, and thus helps analyze what happens when actors form climate-security nexuses.
- By focusing on the interactions between actors without assuming driving forces and structures a priori, translation-based approaches are sensitive to new and emerging phenomena – such as the climate-security nexus – making them helpful for contexts and processes in which established routines and institutionalizations do not exist.
- Finally, the approaches enable avoiding the limitations posed by assuming a strict a priori dichotomy between "science" and "policy", which would prejudge the empirical observations about the role of knowledge resources in the formation of the climate-security nexus.

Alternatives

Options for studying knowledge and climate-security include the *securitization* approach (Wæver 1997; Buzan et al. 1998; Trombetta 2008; Diez et al. 2016), which considers *securitization* as a process that frames issues as security issues within a discourse. A *securitization move* happens when actors make *speech acts* to convince an *audience* that a *referent object* is under an *existential threat*, and thus to gain support for *extraordinary policies* to protect it. Success depends on how the moves are made and received. *Actors* are groups or individuals making moves. A *referent object* is anything presented as threatened. An *audience* is a group or individual whose consent is required to legitimize extraordinary policies. The securitization approach focuses on discourses (Buzan et al. 1998; Aradau 2010; Huysmans 2011), considering securitization as an effect of a "grammatical connection between the threat and the referent object, which itself is accepted by an unspecified audience" (Salter 2019, 353). It helps understand speech acts that link climate and security, actors, and adoption of extraordinary measures.

As outlined in Chapter 2, securitization has been the "go-to" approach for environmental security. Buzan et al. described the nascent efforts to securitize the environment and climate change, and recognized the role of scientific arguments in identifying threats: when governments try to reduce uncertainty, they increase the power of epistemic communities (1998, 71). They observed that "the extent to which scientific argument structures environmental security debates [is] exceptional" (Ibid., 72–3). They also considered that by investigating issues, influencing agendas and communicating to public and decisionmakers, the scientific community often leads environmental securitization (Ibid., 77) and a decisive factor is whether states and others "embrace the scientific agenda" (Ibid., 91). Berling has investigated the role of science in environmental securitization, arguing that scientific objectivism enables science to fix realities, close debates, determine status of actors, mobilize facts and prescribe actions (2011, 390–3). Due to the special status of science, it can "influence on what is being said and what not" (Ibid., 394). But she notes that the ways in which science influences securitization are "yet to be developed to the full" (Ibid., 387). Similarly, Mayer observes that "the role that science plays when climatic threats are discursively anchored is underconceptualized" (2012, 167).

The sociology of translations can complement the securitization approach by focusing on the role of the sciences, and considering situations where securitization is incomplete and controversial. It helps document the "pre-discursive" inputs – that is, knowledge resources – that enable securitizing speech acts, and the interactions of discursive and non-discursive aspects of climate-security debates. It enhances the focus on practices, methods, representations, documents and materials on which discourses depend (which Huysmans (2011) refers to as "little security nothings"). Salter has highlighted the need for a "space for the material and semiotic resources to play a role in the credibility of particular claims about facts, evidence, or reality" in securitization (Salter 2019, 355). The approach also generates insights about the changing key actors. Rychnovská et al. suggested that expertise in security governance is evolving, and that "highlighting and analyzing the different constellations and practices of scientific security expertise is thus crucial for understanding the politics in novel security arenas" (2017, 330).

Another option would be practice theory, often used to consider the role of knowledge in IR (Schatzki 2001; Friedrichs and Kratochwil 2009; Lederer 2012; Bueger and Gadinger 2018). These diverse approaches consider activities and routines conducted by people as generative of international politics. The interest is in (a) what people are actually doing in international politics; (b) how structures evolve through activities; (c) how knowledge is embedded in practices; (d) abandoning efforts to explain via grand theory, and rather describing as in cultural studies and ethnography (Lederer 2012, 641–4). From this perspective, climate-security could be approached through Bourdieu's *praxeology* by conceptualizing it as a *field* – a new space where new practices are negotiated and established – and the climate-security discourse as a competition for position on this field through the accumulation of *capital* (Bueger and Gadinger 2018, 43–4). Climate-security links could then be studied by mapping the *habitus* of the actors involved. This is helpful in situations where the field has been established, and practices are stable, regular, and repetitive. However, the field of climate-security is unstable and contested, with no established practices or habitus. Translations-based approaches help in situations of controversy, uncertainty and improvisation, such as the merging of climate and security. This means describing the climate-security nexus "in the making" (Latour 1987, 12). They provide tools to document the role of knowledge and technologies, which are essential for the "vast machine" of climate sciences (Edwards 2013), as well as enable focus on associations and relations between actors, which are a key consideration when assessing the formation of a nexus.

Guiding concepts

To operationalize the sociology of translations, this study adopts a set of concepts and principles associated with ANT, and adapted to the aims of this study. The five guiding concepts are *translation*, *actor*, *knowledge resources*, *actor-networks* and *black boxes*.

The work of actors to associate climate and security is conceptualized as *translations*. Callon described translation as the mechanism that forms social and natural worlds (1986, 213). Translations integrate separate entities by establishing specific connections between them (Best and Walters 2013(a)). Under ANT's relational ontology, entities acquire social meaning, identities and roles in relation to others (Bueger and Stockbruegger 2017). Everything that has social existence is a result of web of relations between actors connected through translations (Barry 2013, 414). Translation happens when actors associate previously disconnected entities through any activity (such as a speech act, a method or a publication). Through translations, entities are integrated into an intelligible relationship – an *actor-network* – and start functioning in a quasi-coordinated fashion, gain comparability, rules of interaction, and new identities through changed relations with others (Callon 1986; Law 2009; Best and Walters 2013(a)). Translations leave traces in the form of messages, documents and other communication materials that enable monitoring them (Latour 2005, 242). The concept of translation helps when existing categories are difficult to apply. For example, when innovations proliferate, boundaries are uncertain, entities fluctuate, controversies arise and new phenomena emerge (Ibid., 6). So this book considers the formation of the climate-security nexus as a process of translation where two disconnected areas are unified by collective translations. This enables tackling the two key characteristic of climate-security links: the process of merging, and the role of the knowledge. Within this approach, climate-security nexus is not something with an independent existence to be discovered. Rather, it is considered to be a consequence, not a cause, of associations built via translations, and it can be analyzed by documenting empirically the associations that actors build when they connect climate and conflicts (Ibid., 238).

Second, *actors* are any entities, either human or nonhuman, that participate in the translations by making a difference (Latour 2005; Sismondo 2008; Mol 2010). Like all social things, *actors* are a result of webs of relations (Barry 2013, 414; Braun et al. 2019, 787). According to Callon, an "actor does not exist outside the relationships which he enters" (1986, 210). For Latour, an actor is "not a source of action but the moving target of a vast array of entities swarming towards it" (Latour 2005, 46). It is not just an autonomous, conscious and rational individual (although it can achieve that attribute).[2] In Mol's words, "against the implied fantasy of a masterful, separate actor, what is highlighted is the activity of all the associated actors involved. A strategist may be inventive, but nobody acts alone" (2010, 256). The influences and identities of actors are the sum of the information circulating between them, or, in other words, the fields of relations into which they enter (Barry 2013, 414). And a translation is a process of negotiating the identities, groupings and interests of actors (Callon 1986). Anything that makes a difference qualifies as an actor (Latour 2005; Mol 2010; Bueger and Stockbruegger 2017). This means that descriptions of social phenomena should consider multiple types of actors, including humans or nonhumans (Callon 2007, 273). There are two reasons for this: First, human and nonhuman actors influence each other: "Technologies are not passive . . ., they actively intervene in the situations in which they are put to use" (Hogle 2008, 845). So action is a result of associations between humans and nonhumans (Latour 2005, 44). Second, technologies and nonhuman systems are essential for social continuity because they allow recording, communicating and maintaining the resources and standards that enable intersubjectivity – they are the infrastructure of social phenomena (see Aradau 2010).

In this study, *knowledge resources* are understood as anything that provides information that enables translating climate and conflicts together. This refers to both the information that the actors assemble (e.g. interview statements, rainfall data, citations) and the entities involved in the generation, circulation and modification information (measurement devices, databases, statistical methods and indices). "Resources" should not be understood as a passive pool of information that the actors draw upon. Rather, knowledge resources are actors in their own right, since they make a difference in the state of affairs. However, in this study, they are labeled "resources" and considered to be different from the contributions because they themselves do not directly address questions of climate-conflict links. Instead, they provide the information basis which other actors then translate into a set of representations within the climate-conflict debates.

Translations, including the mobilization of knowledge resources, can result in *actor-networks*. The term originates with Callon, who used it "to describe the network of constraints and resources that results from a series of operations of translation" (1986, 221). Under relational ontology, an actor-network is a set of relations, which connects actors, creates identities and capacities to act, and constraints and enables behavior (Mol 2010, 258). Second, it can be understood as a network-shaped description of influences between actors, constructed by the analyst of traces left by translations (Latour 2005, 132). In both meanings, the actor-network is not a thing out in the world to be described. Rather, it is (a) a sum of associations between actors; (b) a tool to describe things by illustrating the associations between actors[3] and the principles of their organization. Depending on the nature and number of associations, actor-networks can be stable, fluid (Mol 2010, 260), fragile, provisional and subject to change. Law describes them as "the provisional assembly of productive, heterogeneous, and (this is the crucial point) quite limited forms of ordering located in no larger overall order" (2009, 146). They determine possibilities for action on the basis of what Latour and Mol call "modes of ordering": they create constraints and incentives that influence actions that follow. Callon describes actor-networks as "constraining", while Mol emphasizes "enabling". To participate in actions, actors need to pass through certain "obligatory passage points" determined by the structure of the actor-network (Callon 1986). By this constraining and enabling, the actor-networks influence actors (Best and Walters 2013(a), 333). Finally, as indicated earlier, actor-networks can include humans and nonhumans – anything that makes a difference in a course of actions and/or the structure of the actor-network.

In addition, translations might turn entities into what Latour calls a "black box". This means that, if translations achieve consensus about an entity, its history, inner workings and networks that hold it together will disappear from view. This happens when translations are deemed beyond question, and the margins of maneuver of entities will become limited by the black box. Initial problematization of issues tends to define negotiable hypotheses about the entity, but translations might end when "a constraining network of relationships has been built" around hypotheses (Callon 1986). A thing might become "unproblematic and certain", a routine assumption, and thus an unquestionable basis for action (Latour 1987, 3). The establishment of black-boxed facts requires collective efforts by actors to subject facts to trials and thus solidifying them. Often such solidification requires that controversies are settled (Ibid., 11). In context of climate and security, many actors attempt to solidify climate-security into a black box. However, "things never unfold quite as planned" (Best and Walters 2013(a), 333). Instead of allowing things to turn into black boxes by accepting and reproducing statements about them, actors might question the conditions of production of such statements (Latour 1987, 29) and challenge previous translations (Callon 1986, 211). Dissent can thus challenge previous translations (Ibid., 213). Such moves can prevent *black boxes* from forming and lead to controversies – as happens, for example, when actors consider climate-conflict links. And even after translations have generated fixed identities and relationships, any *black box* can be reopened by questioning its origins and mobilizing other actors. This way, the fate of translations depends on what others do with them – they might move toward being a "fact", or criticism might do the opposite (Latour 1987, 38). From either perspective, statements need connections to gain momentum and credibility. This way, the production of identities, facts and rules of interaction is a collective process.

Guiding principles

The main principle of analyzing translations is to "follow the actors" (Callon 1986; Latour 2005; Best and Walters 2013(b), 346). Latour argues that "defining and ordering the social is done by the actors, not the analyst" (Latour 2005, 23). This means documenting and describing how the actors themselves translate climate and security together. This book describes how actors articulate the associations between the two, and with what knowledge resources. The descriptions are based on the traces of translations (Latour 2005). To trace how defining and ordering happening, the approach is to document what and how actors are doing, saying, theorizing and communicating. The objective is to follow "the actors in order to identify the manner in which these define and associate the different elements by which they build and explain their world, whether it be social or natural" (Callon 1986, 199). For this book, this means identifying the actors involved in linking climate change and the conflicts in Darfur and Syria, mapping the associations the actors have established through translations, finding adequate ways to illustrate those associations, and through such mapping scrutinizing the content of what is being assembled under the aegis of climate and security (Latour 2005, 5).

The second principle is to start with controversies. Situations of controversy are important because they force actors to translate more intensely and thus build new associations and evidence (Callon 1986; Latour 2005). Associations are traced by following the translations done by actors to stabilize controversies. This enables mapping the connections without trying to settle the controversy or introduce coherence and consistency to the data (Latour 2005, 15–16, 23–4). The climate-security discourses about Darfur and Syria involve various controversies about whether the two were climate wars, about the mechanisms linking climate and the conflicts, and about related factors (such as the implications of climate-conflict links beyond the conflicts themselves, unintended consequences of the links, and underlying understandings of causality).

The third principle is to avoid a priori theoretical assumptions about what populates the world and drives actions (Latour 2005; Irwin 2008; Nexon and Pouliot 2013; Bueger and Stockbruegger 2017; Braun et al. 2019). Such assumptions could normalize certain types of relationships, distract from how the actors themselves translate and articulate the climate-security links, and gain explanatory power. Latour considers that researchers often assume, before data, the existence of "all-purpose social forces" such as a "society", "markets" or "power", which become standard vocabulary, gain the status of causal forces, and are used to explain things (Latour 2005, 22). But summoning them degrades actors to mere informants about such forces (Ibid, 4). Braun et al. have observed how IR research often considers agency and actors as something to be determined prior to empirical research, at the risk that the analysis will find what is theoretically expected (2019, 790). Imposing such concepts can prevent insight into what constitutes phenomena – it "causes you to invent more artifacts and miss the real entities" (Latour 2005, 240–1). Braun et al. have observed that

> if we "temporarily" resolve the question of agency by theoretical assertion before actually looking at what we are interested in, we have already decided what kind of world we want to see: a world of states, say, or a world of individuals, or a world of competing bureaucracies.

> (Braun et al. 2019, 791).

Instead, this study aimed at reconstructing social processes based on observations, not determining them a priori (Nexon and Pouliot 2013, 344). For ANT, social phenomena are consequences, not causes, of the work by actors (Latour 2005, 238), and the analytical task is to inspect the ingredients of social aggregates, such as "security", based on the associations between actors (Ibid., 8), without "purifying" the discussion with predetermined concepts (Irwin 2008, 592). Concepts such as "security" or "climate-conflict links" are not the glue that binds – they are rather a result of associations of specific connectors built through translations (Latour 2005, 5). With this in mind, this book strives to avoid making a priori assumptions about the components of the climate-security nexus or entities that are central to the climate-security discourses, categorize the participating actors into familiar categories or typologies, approach them as representatives of theoretically deduced "social" forces, predetermine what statements are acceptable, speculate about the normative goals of actors, provide "automatic explanations" for the behavior of things (Ibid., 16, 23, 220–1) or suggest how complexities of associations could be simplified. Instead of starting the analysis by positing the existence of aggregate concepts and using them to explain others, this analysis takes them as part of the puzzle (Ibid., 5).

Obviously, descriptions will not give a "full" account of climate-security links. After a mapping, there will remain a great amount of things relevant to climate-security links that cannot be documented because actors have not generated sufficient public knowledge of them, a single book cannot consider all interactions, or because they have become black boxes hidden within complex systems (Ibid., 201). Accounts will thus remain "incomplete, open-ended, hesitant". This is an inevitable consequence of focusing on details of interactions (Ibid., 243). While this "area" remains out of sight, it does influence actors, provide resources for them, and contribute to the generation of climate-security links (Ibid., 244–5). When mapping the translations by actors to link climate change and security, this uncertainty must be accepted, but the study can seek to provide an adequate description of the actor-networks, and orientation for how future research can delve further to describe the climate-security actor-networks.

The study does not aim at strict methodology and replicability, try to access the world by a "disinterested gaze" or aim at writing reports in an "objective style". Latour argues that all accounts are artificial (= man-made), but this is not the opposite of rigor and objectivity, which can be promoted by:

- Maximizing attention to the objects in the world – understanding "objectivity" not as an artificial act of detachment but as a systematic effort to stay focused on the objects that one tries to describe (Ibid., 124–5);
- Strict empiricism, which means describing as far as possible only what the actors themselves are providing (Ibid., 236). This means capturing the actors' own expressions – the concepts of the actors should be stronger than those of the analyst (Ibid., 30);
- Documentation of steps taken during the analysis (Ibid., 133);
- Exchange, meaning that the analyst must aim, to the extent possible, to answer all objections that could be raised against the study (Ibid., 47).

Empirical basis: contributions to "climate war" – debates

This book outlines how the role of knowledge resources in the formation of the climate-security nexus was studied by describing how actors articulate climate-conflict links in context of two conflicts often characterized as climate wars: the 2003–2005 civil war in the Darfur region of Sudan, and the civil conflict in Syria that began in 2011. These two were selected because of their central role in the climate-security debate. While the links between climate and conflict have been explored in since the 1990s, and while many conflicts have been connected by climate change, Darfur and Syria are different because they are paradigmatic examples of climate conflicts.

Other conflicts described as being influenced or triggered by climate change include those in the Lake Chad region (Okpara et al. 2016; Nagarajan et al. 2018; Vivekananda et al. 2019), Somalia (US Ambassador Rice, in UNSC 2011) and Yemen (Douglas 2016). Specific regions, including the Arctic (IPCC 2022, 7–63), Sub-Saharan Africa (Burke et al. 2009), and Bangladesh (Day and Caus 2020; IPCC 2022, 7–81) have been identified as potential areas of climate-induced tensions. Climate change has also been connected, though not necessarily as a causal factor, with the 2022 Russian invasion of Ukraine (Sikorsky et al. 2022).

However, no other conflict has received as much attention as Darfur and Syria. In these cases, existing major conflicts – in fact, possibly the two deadliest conflicts of the 21st century – have been attributed to climate change, and are by far the most often cited examples of climate-conflict links. These are the two instances when a major war at the center of global attention has been systematically connected with by climate change. Darfur is important because it is the first conflict to be systematically labelled as a "climate war" (see e.g. Sachs 2004 (in Srinivasan and Watson 2013); Mjøs 2007; Welzer 2012), and because of the special attention it has received in research and policy debates. Darfur's centrality is illustrated by the fact that the IPCC specifically reviewed climate-Darfur links in its fifth assessment (IPCC 2014). Syria is the most referenced example of a climate conflict, and its central role in the climate-conflict debate is underlined by the IPCC's sixth assessment report (AR6), which reviewed the 20 peer-reviewed studies relating to climate-Syria links (IPCC 2022, 7, 61–2, 16–23).

These two conflicts are the milestones through which the climate-security debate has changed from a general discussion about resources and development into one about the causes extremely destructive wars. Darfur and Syria have brought the climate-security nexus into the heart of the global security discourse. In addition, because these two conflicts involve controversial and extended debates, they enable focusing on controversies, which is central to the research strategy of this book. Given the special attention to and controversies involved in these two conflicts, studying them provides insights that help understand the broader climate-security debate, the climate-security nexus and the role of knowledge within it. As highlighted in the introduction, this book does not aim to cover the broader climate-security debate. Rather, it aims to enhance clarity about that debate and the climate-security nexus by looking at one specific driver: the way in which specific conflicts have been presented as "climate wars". The "climate war" debate provides a helpful window to understand the climate-security debate and nexus because that is where some of the most significant controversies and translations have occurred.

How will these two conflicts be studied? Translations can be monitored based on the traces left by the actors when they work to associate things together (Latour 2005, 8). Such traces are contained in documents that link climate change and the two wars. Therefore, this study will map the translations and knowledge resources that link climate change and the Darfur and Syria conflicts on the basis of a comprehensive set of public documents (speeches, news reports, studies in peer-reviewed journals, reports by governments as well as international and non-governmental organizations, blog entries, and other public communications) which contributed to the debate in context of the two conflicts. For Darfur, the analysis covered 36 documents, and for Syria 35. While several new contributions were published during the writing of this book, and not all could be integrated, the current sample does reflect the main controversies on the climate-conflict links. The documents were selected based on a comprehensive review of the climate-Darfur and climate-Syria debates, and by tracing the references between publications.

Operationalizing the approach

This book describes the Darfur and Syria conflicts as contrasting case studies. This will enable mapping the translations and knowledge resources that drive the linking of climate and the two conflicts, comparing developments across times and regions, describing possible similarities and differences between the two, and considering whether the two enable identifying broader generalizations. Based on the public documents, this study involved four analytical steps:

- Chapters 5 and 6 outline, for Darfur and Syria, respectively, key controversies within the two climate-conflict debates. These are statements which lead to critical reactions.
- For each controversy, the chapters describe how actors undertake translations to settle them by articulating statements about climate-conflict links. This step will describe how actors present entities as connected and outline their identities and roles. This helps understand how new associations produce and reshuffle identities of actors within the climate-conflict debates.
- The third step is to document, for each controversy, the knowledge resources assembled as part of the translations. This means documenting the knowledge that enables actors to represent and debate climate-conflict links and try to settle the controversies. This helps understand the building blocks of relations, the role of materials, technologies and practices; and how the actor-network changes with new resources. The resources will be traced by documenting how the actors describe sources and methods, and by following cross-references.
- On the basis of the first three steps, this book will describe the actor-networks resulting from the translations, thereby considering which actors are involved,

how they are connected, and how the actor-networks influence future activities of others – in other words, how they generate power and influence within the debates. On that basis, this study will describe how knowledge resources feature within the actor-network, what modalities of translation are predominant and how patterns of influence evolve within the climate-conflict debates.

Such a mapping will enable a systematic description of how actors connect climate and the conflicts, the knowledge resources involved, and the patterns of influence emerging from the associations. The mapping will be captured in a written account, which, because of the guiding principle of following the actors, aims to replicate and express how the components of a social phenomenon circulate and become visible in the captured network of associations (Latour 2005, 136). The account should be adapted to the activities of the actors, and make visible the traces left by them, relations between them, and the modes of ordering (Ibid., 128–33). It should express how climate-conflict links become visible in the network of associations, thus "performing" this key step in the development of the climate-security nexus by giving it an artificial textual form that helps understand novel aspects of it. The text should, in particular, describe an increased number of actors involved, agencies that make actors act and objects that stabilize groups and agencies (Ibid., 136–8).

Thus, this book is not a "full" account of the formation of the climate-security nexus, the broader climate-security debate, or even the making of the two climate wars. Rather, it describes the translations of the actors, followed by considerations of the actor-network into which they are organized, while acknowledging that the public documents are just the tip of the iceberg, and that the bulk of the work to form climate-conflict links happens in closed offices, discussions, data centers, and on the desktops and minds of committed actors. This book also does not try to clarify whether climate and security are really connected, whether Darfur and Syria can be characterized as "climate wars", what motivates actors to link climate and security, whether the climate-conflict debates reflect behaviors expected under theoretical frameworks, who "holds" or does not "hold" power in the debates, or whether the actors conform to ideal categories. The conclusions of this book will not confirm or question whether climate change does or does not impact security. It obviously does. But there are many ways to connect the two, and it is important to understand the different approaches to ensure that responses are well-informed and not based on limited information or hasty conclusions.

Limitations

While the sociology of translations adopted for this book enables describing the role of knowledge resources in the translations that form the climate-conflict link, there are limitations relating to the reduced attention to macro-level concepts, human agency, and questions of power (Best and Walters 2013(a), 334). For instance, translation-based approaches do not directly address IRs key questions such as power, the structure of the international system, actions of nation-states, anarchy, or the agent-structure problem. Rather, the results are situated and case-specific, and do not necessarily allow testing or extracting generalizable laws of behavior of actors within the international system. Nexon and Pouliot observed that ANT lacks ways to capture macro phenomena, which can be challenging from the IR perspective, where phenomena often have pronounced macro dimensions (2013, 344). On a related note, Erickson (2016) argues that ANT merely encourages relabeling of things as translations between mediators in a network but does not clarify why things are happening. And this lack of explanatory power is strengthened by refusal apply "generally accepted ontological categories". However, as will be demonstrated in this study, certain patterns of macro-level dynamics can be discerned by identifying the modes of ordering of the actor-networks.

Another common criticism is the neglect of human agency. Comparing ANT with practice theory, Lederer considers that ANT tends to lose focus on forms of power that depend on humans (2012, 652). Collins (2010) argues that treating nonhumans as intelligent participants in social knowledge misrepresents how the world works because only humans possess language and the capacity to interpret – two criteria for being "social". For Collins, nonhuman actors cannot operate without human minders, and ANTs symmetry between humans and nonhumans ignores this. Furthermore, he considers that the outputs of scientific instruments are subject to human interpretations, concluding that "never treat an instrument as an intelligent actor or a participant in social knowledge unless you have a very good methodological reason for doing it" (2010, 146).

These are an important reminder that, to provide comprehensive descriptions, accounts need to remain agnostic about what types of actors – whether human or nonhuman – are influential. The studies described in this book indicate that, while technical actors are essential for developing, maintaining and circulating the knowledge basis for the climate-conflict links, human actors might be essential for articulating coherent narratives about the nexus. Thus, rather than prioritizing one over the other, this book aims to consider the different forms of influence of a wide range of actors.

Aradau highlighted that ANT tends to leave out political history and the impact of performativity on issues of gender (2010, 496). Barry (2013), described how feminist and post-colonial authors have criticized ANT's tendency to adopt the perspective of the empire rather than the subjects (415). And Star (1991) argued that ANT tends to be centered, managerialist, and militaristic, attending to the powerful in a functionalist and masculinist mode. Best and Walters note that ANT does not consider domination and inequality, and argue that such processes are an unfortunate part of international politics and thus IR research[4] (2013(b), 346).

The list of contextual aspects that ANT-based approaches might miss includes norms (Shapin 1998), culture, power (Star 1991; Lederer 2012; Erickson 2016), gender, "otherness" (Law 2004) and politics (Anderson and Adams 2008). Due to this limited attention to context, Erickson concludes that ANT stories can get bogged down in details of local networks, "might not tell us everything we need to know" and fail to challenge power imbalances in networks (2016, 94). Similarly, Hess considers that "discussions of actor-networks need to be framed by an analysis of culture and power" and used to illustrate how structural change is possible (1995, 53).

Finally, a more technical limitation is that IR issues, especially those related to national security, involve high levels of secrecy. Therefore, in many areas it is not possible to follow the actors, although Best and Walters do suggest that it is possible to follow those who expose secretive security practices (2013(b), 346).

Despite these limitations, and as acknowledged also by several IR and security studies scholars, the translation-based approach can generate new insights into how areas of governance emerge or are merged into nexuses, how knowledge resources are mobilized to generate such areas, and how the actor-networks create new constellations of power and influence. Thus, the sociology of translations can enrich the landscape of empirical studies of climate-conflict links.

Notes

1 The term ANT has stuck, though Callon originally suggested "sociology of translations" (1986), Law preferred "material semiotics" (2009), and Latour regrets the term because it is easily confused with network theory (2005).
2 A classic example is Latour's study of Pasteur and the anthrax vaccine (1983). Pasteur is generally considered a "great man". However, Latour describes how his discovery involved many other people (journalists, farmers and vets) and things (blood, cows, bacteria, laboratory equipment). However, narratives of great deeds tend to leave out the heterogeneous actors over time.
3 Latour uses the metaphor of a "network" because nets involve: (a) traceable and empirically recordable point-to-point connections between sites; (b) empty spaces – like the aspects of social phenomena that cannot be observed; (c) effort – they require work by actors to be created (Latour 2005, 131–132).
4 Law (1986), studied the socio-material networks that enabled the Portuguese maritime empire, and Austin's two studies on torture (2015, 2016) use ANT-based approaches to document the material-communicative networks that generate the global phenomena of torture.

Literature

Anderson, W. and Adams, V. 2008: *Postcolonial studies in technoscience*. In: Hackett et al. 2008, 181–200.

Aradau, C. 2010: *Security that matters: Critical infrastructure and objects of protection*. In: Security Dialogue, 41(5), 491–514.

Austin, J.L. 2015: *We have never been civilized: Torture and the materiality of world political binaries*. In: European Journal of International Relations, 23(1), 49–73.

Austin, J.L. 2016: *Torture and the material-semiotic networks of violence across borders*. In: International Political Sociology, 10, 3–21.

Barry, A. 2013: *The translation zone: Between actor-network theory and international relations*. In: Millennium: Journal of International Studies, 41(3), 413–29.

Berling, T.V. 2011: *Science and securitization: Objectivation, the authority of the speaker and mobilization of scientific facts*. In: Security Dialogue, 42(4–5), 385–97.

Best, J. and Walters, W. 2013(a): *Actor-network theory and international relationality: Lost (and found) in translation*. In: International Political Sociology, 7(3), 332–4.

Best, J. and Walters, W. 2013(b): *Translating the sociology of translations*. In: International Political Sociology, 7(3), 345–9.

Bijker, W.E.; Bal, R. and Hendriks, R. 2009: *The Paradox of Scientific Authority: The Role of Scientific Advice in Democracies*. MIT Press, Cambridge, MA.

Boswell, C. 2009: *The Political Uses of Expert Knowledge: Immigration Policy and Social Research*. Cambridge University Press, Cambridge.

Braun, B.; Schindler, S. and Wille, T. 2019: *Rethinking agency in international relations: Performativity, performances and actor-networks*. In: Journal of International Relations and Development, 22, 787–807.

Bueger, C. and Gadinger, F. 2018: *International Practice Theory*. Palgrave Macmillan, Cham.

Bueger, C. and Stockbruegger, J. 2017: *Actor-network theory: Objects and actants, networks and narratives*. In: McCarthy, D. (Ed.): Technology and World Politics. Routledge, Abingdon, 42–59.

Burke, M.B.; Miguel, E.; Satyanath, S.; Dykema, J.A. and Lobell, D.B. 2009: *Warming increases the risk of civil war in Africa*. In: PNAS, 106(49), 20670–4.

Buzan, B.; Waever, O. and de Wilde, J. 1998: *Security – A New Framework for Analysis*. Lynne Rienner Publishers Inc., London.

Callon, M. 1986: *Some elements of a sociology of translation: Domestication of the scallops and the fishermen of St. Brieuc Bay*. In: Law, J. (Ed.): Power, Action and Belief: A New Sociology of Knowledge? London, Routledge, 196–223.

Callon, M. 2007: *Actor-network theory – the market test*. In: Asdal, K.; Brenna, B. and Moser, I. (Eds.): Technoscience – The Politics of Interventions. Academic Press, Unipub Norway, Oslo, 273–86.

Collins, H. 2010: *Humans not instruments*. In: Spontaneous Generations: A Journal for the History and Philosophy of Science, 4(1), 138–47.

Day, A. and Caus, J. 2020: *Conflict Prevention in an Era of Climate Change: Adapting the UN to Climate-Security Risks*. United Nations University, New York.

Diez, T.; von Lucke, F. and Wellmann, Z. 2016: *The Securitisation of Climate Change: Actors, Processes and Consequences*. Routledge Prio New Security Studies. Routledge, Oxon.

Douglas, C. 2016: *A Storm Without Rain: Yemen, Water, Climate Change and Conflict*. 3 August. At: https://climateandsecurity.org/2016/08/a-storm-without-rain-yemen-water-climate-change-and-conflict/ (accessed 24 July 2022).

Edwards, P.N. 2013: *A Vast Machine – Computer Models, Climate Data, and the Politics of Global Warming*. MIT Press, Cambridge, MA.

Erickson, M. 2016: *Science, Culture and Society: Understanding Science in the Twenty-First Century*. Polity Press, Cambridge.

Friedrichs, J. and Kratochwil, F. 2009: *On acting and knowing: How pragmatism can advance international relations research and methodology*. In: International Organization, 63, 701–31.

Fuller, S. 2006: *The Philosophy of Science and Technology Studies*. Routledge, New York.

Henriksen, L.F. 2013: *Performativity and the politics of equipping for calculation: Constructing a global market for microfinance*. In: International Political Sociology, 7, 406–25.

Hess, D.J. 1995: *Science and Technology in a Multicultural World: The Cultural Politics of Facts and Artifacts*. Columbia University Press, New York.

Hogle, L.F. 2008: *Emerging medical technologies*. In: Hackett et al. 2008, 841–74.

Huysmans, J. 2011: *What's in an act? On security speech acts and little security nothings*. In: Security Dialogue, 42(4–5), 371–83.

IPCC 2014 (Adger, W.N.; Pulhin, J.M.; Barnett, J.; Dabelko, G.D.; Hovelsrud, G.K.; Levy, M.; Oswald Spring, Ú. and Vogel, C.H.): *Human security*. In: Field, C.B.; Barros, V.R.; Dokken, D.J.; Mach, K.J.; Mastrandrea, M.D.; Bilir, T.E.; Chatterjee, M.; Ebi, K.L.; Estrada, Y.O.; Genova, R.C.; Girma, B.; Kissel, E.S.; Levy, A.N.; MacCracken, S.; Mastrandrea, P.R. and White, L.L. (Eds.): Climate Change 2014: Impacts, Adaptation, and Vulnerability. Part A: Global and Sectoral Aspects. Contribution of Working Group II to the Fifth Assessment Report of the IPCC. Cambridge University Press, Cambridge and New York, NY, 755–91.

IPCC 2022 (H.-O. Pörtner, D.C. Roberts, M. Tignor, E.S. Poloczanska, K. Mintenbeck, A. Alegría, M. Craig, S. Langsdorf, S. Löschke, V. Möller, A. Okem, B. Rama (Eds.)): *Climate Change 2022: Impacts, Adaptation, and Vulnerability. Contribution of Working Group II to the Sixth Assessment Report of the IPCC.* Cambridge University Press. In Press. At: www.ipcc.ch/report/sixth-assessment-report-working-group-ii/ (accessed 19 June 2022).

Irwin, A. 2008: *STS perspectives on scientific governance.* In: Hackett et al. 2008, 583–608.

Jasanoff, S. (Ed.) 2004: *States of Knowledge – The Co-Production of Science and Social Order.* Routledge, Oxon.

Latour, B. 1983: *Give me a laboratory and I will raise the world.* In: Knorr-Cetina, K.D. and Mulkay, M. (Eds.): Science Observed – Perspectives on the Social Study of Science. Sage Publications, London, 141–70.

Latour, B. 1987: *Science in Action – How to Follow Scientists and Engineers Through Society.* Harvard University Press, Cambridge, MA.

Latour, B. 2004: *The Politics of Nature – How to Bring the Sciences into Democracy.* Harvard University Press, Cambridge.

Latour, B. 2005: *Reassembling the Social – an Introduction to Actor-Network Theory.* Oxford University Press, Oxford.

Law, J. 1986: *On the methods of long distance control: Vessels, navigation, and the Portuguese route to India.* In: Law, J. (Ed.): Power, Action and Belief: A New Sociology of Knowledge? Sociological Review Monograph 32. Routledge, Henley, 234–63.

Law, J. 2004: *After Method: Mess in Social Science Research.* Routledge, London.

Law, J. 2009: *Actor network theory and material semiotics.* In: Turner, B. (Ed.): The New Blackwell Companion to Social Theory. Blackwell Publishing, Hoboken, NJ, 141–58.

Lederer, M. 2012: *The practice of carbon markets.* In: Environmental Politics, 21(4), 640–56.

Mayer, M. 2012: *Chaotic climate change and security.* In: International Political Sociology, 6, 165–85.

Mjøs, O.D. 2007: *Nobel Peace Prize 2007 Award Ceremony Speech.* The Nobel Foundation. 10 December. At: www.nobelprize.org/prizes/peace/2007/ceremony-speech/ (accessed 9 November 2020).

Mol, A. 2010: *Actor-network theory: Sensitive terms and enduring tensions.* In: Kölner Zeitschrift für Soziologie und Sozialpsychologie, Sonderheft, 50, 253–69.

Nagarajan, C.; Pohl, B.; Rüttinger, L.; Sylvestre, F.; Vivekananda, J.; Wall, M. and Wolfmaier, S. 2018: *Climate-Fragility Profile: Lake Chad Basin.* Adelphi, Berlin. At: www.adelphi.de/en/system/files/mediathek/bilder/Lake%20Chad%20Climate-Fragility%20Profile%20-%20adelphi_0.pdf (accessed 21 November 2019).

Nexon, D.H. and Pouliot, V. 2013: *Things of networks: Situating ANT in international relations.* In: International Political Sociology, 7(3), 342–5.

Okpara, U.T.; Stringer, L.C. and Dougill, A.J. 2016: *Perspectives on contextual vulnerability in discourses of climate conflict.* In: Earth System Dynamics, 7, 89–102.

Rothe, D. 2017: *Seeing like a satellite: Remote sensing and the ontological politics of environmental security.* In: Security Dialogue, 48(4), 334–53.

Rychnovská, D.; Pasgaard, M. and Berling, T.V. 2017: *Science and security expertise: Authority, knowledge, subjectivity.* In: Geoforum, 84, 327–31.

Salter, M. 2019: *Security actor-network theory: Revitalizing securitization theory with Bruno Latour.* In: Polity, 51(2), 349–64.

Schatzki, T.R. 2001: *Practice mind-ed orders*. In: Knorr-Cetina, K. (Ed.): *The Practice Turn in Contemporary Theory*. Taylor and Francis, London, 50–63.

Shapin, S. 1998: *Placing the view from nowhere: Historical and sociological problems in the location of science*. In: Transactions of the Institute of British Geographers, 23(1), 5–12.

Sikorsky, E.; Barron, E. and Hugh, B. 2022: *Climate, Ecological Security and the Ukraine Crisis: Four Issues to Consider*. CCS Briefer No. 31. 12 March. At: https://climateand-security.org/wp-content/uploads/2022/03/Climate-Ecological-Security-and-the-Ukraine-Crisis_Four-Issues-to-Consider_BRIEFER-31_2022_11_3.pdf (accessed 21 July 2022).

Sismondo, S. 2008: *Science and technology studies and an engaged program*. In: Hackett et al. 2008, 13–31.

Srinivasan, S. and Watson, E.E. 2013: *Climate change and human security in Africa*. In: Redclift, M.R. and Grasso M. (Eds.): *Handbook on Climate Change and Human Security*. Edward Elgar Publishing Ltd., Cheltenham, UK.

Star, S.L. 1991: *Power, technology and the phenomenology of conventions: On being allergic to onions*. In: Law, J. (Ed.): A Sociology of Monsters. Routledge, London, 26–56.

Trombetta, M.J. 2008: *Environmental security and climate change: Analysing the discourse*. In: Cambridge Review of International Affairs, 21(4), 585–602.

UNSC 2011: *Record of the 6587th Meeting*. Document S/PV.6587. At: https://undocs.org/en/S/PV.6587 and https://undocs.org/en/S/PV.6587(Resumption1) (accessed 20 December 2020).

Vivekananda, J.; Wall, M.; Sylvestre, F. and Nagarajan, C. 2019: *Shoring Up Stability: Addressing Climate and Fragility Risks in the Lake Chad Region*. Adelphi, Berlin. At: www.adelphi.de/en/publication/shoring-stability (accessed 23 November 2019).

Wæver, O. 1997: *Concepts of Security*. Copenhagen University Press, Copenhagen.

Weingart, P. 1999: *Scientific Expertise and political accountability: Paradoxes of science in politics*. In: Science and Public Policy, 26(3), 151–61.

Welzer, H. 2012: *Climate Wars: What People Will Be Killed for in the 21st Century*. Polity Press, Cambridge.

5   Darfur

The first "climate war"?

So far, Chapter 2 described climate-security research and policymaking, illustrated the disconnect between the status of knowledge and policymaking and highlighted how empirical studies into the role of knowledge in the formation of climate-conflict links can illustrate a key aspect of the formation of the climate-security nexus. Chapter 3 connected climate-security with the broader trend of nexus formation in IR and called for a research strategy that is sensitive to the role of knowledge and to processes of linking entities into nexuses. Chapter 4 described the research strategy adopted for this book, highlighting how the role of knowledge in the formation of the climate-security nexus can be better understood by looking at how specific climate wars are made, which, in turn, can be studied with a sociology of translations inspired by ANT.

This chapter will describe how actors debate the link between climate change and the Darfur civil war. It begins with a short conflict background, then describes the contributions to the climate-Darfur debate (which were the materials for this analysis). This is followed by the identification of six key controversies, and descriptions of how actors undertake translations to try to settle each of those controversies, and how they mobilize knowledge resources within those translations.

Conflict background

The Darfur civil war that began in 2003 was the first conflict to be described as a "climate war" – a conflict whose outbreak was, to a significant extent, more likely due to anthropogenic climate change. Research has identified many conflict factors, including water and land disputes, ethnic diversity, discrimination, discriminatory colonial structures, Khartoum's brutality and policies of neglect, influx of weapons from Libya, foreign influence (attributed, for example, to Chad, Eritrea, the United States, Britain, South Sudan, and Israel), environmental challenges, population growth, lack of conflict resolution mechanisms, government weakness or strength and even the prohibition of alcohol (see Anderson 2004; UNEP 2007; Flint and De Waal 2008; Mamdani 2009; Hagan and Kaiser 2011; Akasha 2014; Selby and Hoffmann 2014).

While Sudan had seen conflict for decades (see Flint and De Waal 2008; Mamdani 2009; Hagan and Kaiser 2011), and the underlying causes of the conflict are

DOI: 10.4324/9781003451525-5

disputed and deeply ingrained in history and social dynamics (Mamdani 2009), the violence known as the Darfur civil war is "often traced to the government's response to scattered rebel attacks . . . in the early months of 2003" (Hagan and Kaiser 2011). The escalation began in 2001–2002 with the formation of the Sudanese Liberation Movement (SLM) and the Justice and Equality Movement (JEM) (Anderson 2004; Mamdani 2009; Sunga 2011), who accused the government of neglecting Darfur (Akasha 2014, 8).

In 2002, SLM and JEM began attacking police and army outposts, culminating in an assault on El Fasher airport in April 2003, transforming the war into a military conflict (Flint and De Waal 2008, 120–1). After initial reconciliation efforts failed (Mamdani 2009, 253), the government implemented curfews, arrests and assaults on rebel strongholds. But the rebels kept winning most confrontations in 2003 (Ibid., 255 and Flint and De Waal, 122). Constant losses led the government to mobilize its air force, military intelligence as well as the infamous Janjaweed and other ethnic militias (Ibid., 123–8), whose atrocities against civilians triggered a full-scale civil war (Sunga 2011, 66). The Janjaweed were led by Musa Hilal, released from prison in June 2003 and instructed by Khartoum to "change the demography of Darfur and empty it of African tribes" (Flint and De Waal 2008, 128). Consequently, "state-led attacks on food and water massively dislodged Black Africans in Darfur" (Hagan and Kaiser 2011, 1). Particularly intense attacks and massacres happened in June–September 2003 and during December 2003 to April 2004 (Flint and De Waal 2008).

The first cease-fire talks took place in Abeche, Chad, in August 2003 (Mamdani 2009, 253). In March 2004, the UN Coordinator Mukesh Kapila called the crisis "genocide" (followed by a similar declaration by US Congress) (Ibid.). After March–April 2004, the violence receded as the government reduced its operations, and the conflict settled into patterns of irregular fighting between rebels and militias, as well as between rebel groups. In July 2005, an African Union peacekeeping force was deployed (Ibid., 150–66). Several cease-fires and peace talks took place in 2004–2006, including the Darfur Peace Agreement in Abuja in May 2006 (Mamdani 2009). However, the talks struggled to get Darfur's diverse rebel groups to agree, and donor pressures prevented inclusive and deep negotiations, resulting in a flawed agreement (Ibid., 262). Thus, the period involved walk-outs, reorganizations of rebel groups, banditry and rogue rebel groups, and eruptions of violence, though the fighting never reached 2003–2004 levels (Flint and De Waal 2008, 278–80; Mamdani 2009). In July 2007, following pressure from the United States, EU and the UNSC, Sudan agreed to a UN-African Union force in Darfur.

A study has estimated that the conflict killed 300,000 and displaced three million in 2004–2008 (Degomme and Guha-Sapir 2010; Hagan and Kaiser 2011; Sunga 2011). However, the numbers of victims remains contested (Mamdani 2009, 273). The International Criminal Court (ICC) considered members of the Zaghawa, Masalit and Fur ethnic groups as main victims, and the Janjaweed militias and Sudanese leadership as perpetrators and has filed genocide charges against President al-Bashir (Mamdani 2009; Hagan and Kaiser 2011). Further talks in 2010–2011 resulted in a second peace agreement, also not signed by all rebel groups. Thus, while the situation has not escalated, sporadic violence has continued, and fears of return to violence have re-emerged as African Union-United Nations Hybrid Operation in Darfur ended its operations on 31 December 2020. The August 2019 Draft Constitutional Declaration of Sudan required that a peace process leading to an agreement must be undertaken within the first six months of the 39-month transition period to a democratic government.

The climate-Darfur debate

How did this conflict become presented as the first "climate war"? It was first described as such by Columbia University economist Jeffrey Sachs in November 2004 at Oxford University (Srinivasan and Watson 2013, 315). Soon thereafter, in December 2004, the University of Khartoum and University of Peace organized a conference on *Environmental degradation as a cause of conflict in Darfur*, focusing on how dwindling resources and drought led to poverty, migration and the conflict (UPAP and PRI 2006). In 2006, the *Stern Review* highlighted how droughts have been identified as a conflict factor (Stern 2006, 112–13).

Along similar lines, at Chatham House in February 2006, British Secretary of Defense John Reid described Darfur as a conflict that was aggravated by climate-induced lack of water and land and as a warning of climate conflicts, and argued that UK armed forces must prepare for climate impacts and subsequent wars. The speech was reported on 28 February by *The Independent* (Russell and Morris 2006).

Reid's speech resonated in the United States. On 10 March 2006, Professor Michael Klare of Hampshire College, Amherst, published *The Coming Resource Wars* (Klare 2006) on the news website *tompaine.com*. Klare described Reid's remarks, considered how climate change can threaten national security and elaborated on the implications of climate-related security threats, and possible responses. Further, Professor Sachs, in *Ecology and Political Upheaval* in *Scientific American*, described how climate change can cause wars and how Darfur is rooted in an ecological crisis (2006).

Next, Josh Braun published *A Hostile Climate – Did Global Warming Cause a Resource War in Darfur?* in *SEED Magazine*, and online journal, on 2 August 2006 (Braun 2006). Braun recapped the arguments by Reid and Klare, but also highlighted, for the first time in the debate, the role of Khartoum's neglect rather than climate change. On 6 November 2006, Scott Baldauf's piece *Africans Are Already Facing Climate Change* in *Christian Science Monitor* described how climate change as well as water and land scarcity triggered conflicts between farmers and pastoralists and how this will impact Africa in future.

UK policymakers, media and civil society continued steering the debate. At a Labour Party conference in September 2006, Foreign Secretary Beckett stated that desertification and water shortages fueled the conflict (Hughes 2007). In January 2007, David Cameron, leader of the UK opposition, published *A Warmer World Is Ripe for Conflict and Danger* in the *Financial Times*, describing how lack of rain pitted farmers against herders, and how Britain needs to adapt its climate and defense policies. In March 2007, the NGO Tearfund published the report *Darfur: Relief in a Vulnerable Environment*, focusing on the environment and post-conflict recovery, describing how climate models forecast reduced agricultural output, which poses a risk of further violence (Bromwich et al. 2007).

In April 2007, UNSC held its first debate on climate change (2007(a)). This was initiated by Britain and chaired by Foreign Secretary Beckett. She told the BBC that "one of the reasons for the conflict erupting in Sudan . . . was pressure on land and pressure on water. That is something climate change will make much worse" (Hughes 2007). The debate covered many aspects of climate change, but Darfur was not mentioned (UNSC 2007(b)). However, John Ashton, UK special representative for climate change, stated that "the security implications of climate changes are bigger than we thought . . . Their effects can already be seen in Darfur" (Reynolds 2007).

In the US media, Stephan Faris, a journalist, published *The Real Roots of Darfur* in *The Atlantic* in May 2007. He argued that the conflict may have been primarily caused by climate change, described how pastoralists and farmers began fighting due to drought-induced scarcity, and argued that nations with high GHG emissions could be partly responsible. Following Faris, UN Secretary General Ban Ki Moon published *A Climate Culprit in Darfur* in the *Washington Post* in June 2007. He argued that the conflict arose "at least in part from climate change", reflected by a precipitation decline connected with temperature rise of the Indian Ocean, and hence, global warming. Ki Moon elaborated that this forced farmers to close their land from herders, leading to tensions and escalation in 2003.

In June 2007, shortly after Ki Moon, the UN Environment Programme (UNEP), published *Sudan – Post-conflict Environmental Assessment* (UNEP 2007), the first major report considering environment-conflict links in Sudan. While climate change was not a primary topic, it was addressed as an exacerbating factor which contributed to desertification and herders-farmers tensions.

Ki Moon's article and the UNEP report triggered critical responses. In *Is Climate Change the Culprit for Darfur?* published in June 2007 on the website *African Arguments* (De Waal 2007(a)), Alexander De Waal argued that attributing the conflict to climate is simplistic, that no clear evidence links climate and droughts, and that economic factors, government neglect and militarization turned a food and migration crisis into a conflict. He acknowledged that climate impacts can cause disputes but well-governed polities can settle those non-violently. Another critical piece, *Sudan: Climate Change – Only One Cause Among Many for Darfur Conflict*, was published on *Integrated Regional Information Networks* (IRIN) – a humanitarian news portal operated at the time by UN Office for the Coordination of Humanitarian Affairs (UNOCHA) (IRIN 2007). As de Waal, it emphasized the relevance of other conflict drivers and highlighted the risks of whitewashing the Sudanese government through simplistic linking of climate and the conflict.

De Waal's criticisms were taken up by Thomas Homer-Dixon – Professor at the University of Toronto – in *Cause and Effect*, also on *African Arguments* in August 2007. He argued that in complex nonlinear systems involving ecological and human factors, causes are interactive, and thus it is impossible to determine their relative importance. He deemed invalid de Waal's identification of the Sudanese government as the main culprit, and argued that any explanation needs to include climate as a causal factor (Homer-Dixon 2007). Later, De Waal responded that one can prioritize causes by comparing local developments, noting that violence in Darfur was more intense in areas with lower environmental degradation and rainfall decrease, and citing the empirical consistency between militarization and conflict, and elite resource capture (De Waal 2007(b)).

Criticisms were also raised by Idean Salehyan, political scientist at the University of North Texas, in *The New Myth About Climate Change* in August 2007 in *Foreign Policy*. Referring to Ki Moon (2007), CNA (2007) and Schwartz and Randall (2003), he considered that evidence for links between scarcity, environment and conflict is mostly anecdotal; that, regardless of warming, conflicts have decreases globally; and that climate attribution allows governments to distract from their repressive policies. He concluded that "Ki-moon's case about Darfur was music to Khartoum's ears". Indeed, Sudan's ambassador to the UN, Mr. Abdalmahmood Mohamad, argued at Lehigh University in Pennsylvania, that climate change was the primary cause of the conflict (Straw 2007).

The big year for climate 2007 culminated on 10 December, when the Nobel Peace Prize was awarded to the IPCC and Al Gore. Chairman of the Norwegian Nobel Committee Professor Ole Danbolt Mjøs described Darfur as the world's "first climate war", arguing that desertification triggered migrations, which forced groups into conflict. He outlined how the problem has already spread "from the Sudan to Senegal" (Mjøs 2007).

In April 2008, when French President Nicholas Sarkozy hosted the Major Economies Meeting on Climate Change. He used Darfur as an example of how climate change can prompt migration of the poor and trigger wars, of which there might be dozens in the future (Ingham and Hood 2008).

The first commentary on the climate-Darfur link in a peer-reviewed journal was published in August 2008 by Kevane and Gray (2008) from Santa Clara University in California in *Environmental Research Letters*. They investigated what precipitation and conflict data say about the link between drought and the conflict, and challenged the view that the conflict was associated with precipitation decline.

In 2009, Mamdani's *Saviors and Survivors*, while focusing on the underlying social structures and foreign influences on the conflict, highlighted the climate-induced desertification and erosion across the region, which had regularly led to devastating famines and forced pastoralists to move south to find pasture.

Another test of the climate-Darfur link was conducted by Ian Brown (2010) from Stockholm University. In an article in the *International Journal for Remote Sensing*, Brown applied remote sensing images and the Normalized Difference Vegetation Index (NDVI) to illustrate that vegetation levels in Darfur improved in 1981–2006, and argued that this contradicts assertions that eco-scarcity or climate change provoked the conflict.

The next contribution was Jeffrey Mazo's 2010 book *Climate Conflict*. Mazo – at the time a researcher at the International Institute for Strategic Studies – argued in support of the climate-conflict link, compared Kevane and Gray's findings about rainfall-conflict links with other sources and concluded that temperature and conflicts are correlated across Africa.

In July 2011, Sudan's Ambassador Osman described at the UNSC (UNSC 2011(a)) how drought and desertification in 1985 drove farmers and herdsman into fighting. Quoting a saying: "the herdsman would sooner see his son die before his eyes than his cow", he concluded that desertification and drought were the main causes, and emphasized that had the international community helped Sudan address the basic causes it would have saved the 3 bn USD per year spent on the African Union-United Nations Hybrid Operation in Darfur.

In *Development and Change* (2011), Harry Verhoeven argued that the "climate war" label masks human agency as a determinant of scarcity, as well as the role of the government. He considered that Khartoum manipulated the Malthusian climate-conflict narratives in ways that harm local communities, and that, rather than being based on evidence, climate-security debates result from post-1989 search for foreign policy identities and from attention-seeking by progressives. He concluded that "Darfur is not the world's first climate-change conflict and a-political development is not the solution" (702) and called for more critical assessments of causes and consequences, as well as focus on the demand side of resource crises.

In *Klimakriege* ("Climate Wars"), Harald Welzer, professor at University of Flensburg, argued that Darfur was a war by a government against its people, in which climate change played a key role. After the 1984 drought, farmers closed their fields to pastoralists, who had also lost livelihoods, and began fighting for access. This led to migration, population growth contributed to problems, and Khartoum's policies ruined conflict resolution mechanisms. He concluded that ecological conflicts are often perceived as ethnic, even by the participants themselves (who do not see the ecological collapse, but only inter-group violence) (2012, 93–8).

The IPCC, in its fourth assessment report (AR4), specifically considered the climate-Darfur link in box 12-5 in Chapter 12 ("Human Security") of the report of Working Group II (IPCC 2014, 773). It reviewed five peer-reviewed studies, as well as Mazo's 2010 book, and concluded that most existing research agrees that the conflict had many causes and that Darfur had coped well with past droughts.

Another critical contribution was published by Selby and Hoffmann (2014) in *Global Environmental Change*. They criticized the centrality of farmer-pastoralist relations as outdated in light of Darfur's comparably modern economy, and argued that the subsistence peasant stipulated by climate-conflict narratives no longer exists. As main conflict factors, they highlighted uneven land relations and Khartoum-led reforms that undermined traditional conflict resolution mechanisms. In addition, they warned that climate-conflict narratives tend to obscure Khartoum's responsibility.

On 28 October 2015, adelphi – a Berlin-based think tank – published a video *Drought, Migration and Civil War in Darfur* (Adelphi 2015). It described how the war was intermingled with resource conflict caused by droughts. It warned that claims about direct climate-Darfur links should be treated with caution, and outlined a narrative for the conflict: rains declined in 1960–85, culminating in famine in 1984–1985, forcing pastoralists to migrate to areas of farmers, already under demographic pressure. Darfurians grew discontent, but the government responded by arming only one group. Traditional land management systems broke down, and ethnic superiority discourses aggravated things further.

At an uncertain point in time, the GWR published an entry identifying Darfur as the first climate war, referencing rainfall reductions, tensions between "northern Muslims" and "southern Christians/Animists", as well as competition for land and water (GWR 2019).

Finally, regardless of the critical observations of the IPCC report climate-Darfur links reappeared at a Democratic presidential debate in March 2020, when presidential candidate Joe Biden used Darfur as an example of a climate war caused by changing weather and desertification (Biden 2020). Biden's statement triggered criticisms by Breitbart News (Moran 2020), which argued that analysts attribute that conflict to racial tensions and land rights, and that while imprudent land use was the main reason, climate scientists "devised an alternative theory" indicating that climate change was to blame.

Linking climate change and Darfur

As discussed in Chapter 4, translation means associating previously disconnected entities, thereby defining their identities, operating spaces, and interaction possibilities. The contours of the climate-Darfur debate described in the previous section illustrate how actors translate climate and the conflict together. However, the fate of statements depends on what other actors do with them, and translations can become subjects of controversy. The chronology also shows how the climate-Darfur links invite dissenting contributions, which question the initial translations and mobilize new knowledge resources, and thus various controversies emerge. Table 5.1 identifies the main ways in which climate and Darfur have been connected, as well as the corresponding dissenting translations made in the critical contributions. These are the contours of the climate-Darfur debate and provide the structure for this chapter.

Table 5.1 Main controversies about the climate-Darfur link

Initial translation	*Dissenting translation*
Darfur was categorically a climate war, and a first one of its kind.	Darfur was not a climate war – other causes were more important.
Rains failed before the conflict due to climate change.	Rains did not fail prior to the conflict.
Darfur was suffering from climate-induced resource scarcity prior to the conflict.	Natural resources did not decline in Darfur prior to the conflict.
Lack of water and land drove farmers and pastoralists into conflict with each other.	The Darfurian economy is more complex than farmer-pastoralist dynamics.
The Darfur conflict is a "warning sign" about future climate wars across the globe.	There have been no climate conflicts across Africa.
The Sudanese government in Khartoum bears less responsibility for the conflict due to the influence of climate.	The Sudanese government has found an excuse to avoid responsibility for the conflict; uses cli-security discourse to manipulate world opinion.

The following sections will describe each of these controversies by outlining how the proponents of climate-Darfur links articulate the link, how dissenting contributions criticize it, and what knowledge resources both sides mobilize for their translations.

"You heard it from me first: Darfur is the world's first climate war"

Several contributions stated categorically that Darfur was a climate-driven conflict, with some emphasizing that it was the first of its kind. This was expressed in statements such as:

> You heard it from me first, Darfur is the world's first climate change war.
> (Jeffrey Sachs, quoted in Srinivasan and Watson 2013, 315)

> Darfur may well be the first war influenced by climate change.
> (Braun 2006)

> The consequences are most obvious . . . among the poorest of the poor, in Darfur and in large sectors of the Sahel belt, where we have already had the first "climate war".
> (Mjøs 2007)

Less boldly, UNEP stated that "the impact of climate change is considered to be directly related to the conflict" (2007, 60) and that environmental degradation and "climate instability and change" are major underlying causes (Ibid., 329). More recently, the GWR presented Darfur as the "first climate change war" (GWR 2019), and Joe Biden stipulated that "you saw what happened in Darfur with the changing weather patterns and the desertification there. It causes war, it causes great migrations" (Biden 2020).

But others argued that speaking about a climate war is simplistic, and that other drivers, including government mismanagement and militarization, were more important. IRIN called for more nuanced approaches (IRIN 2007), and De Waal argued that Ki Moon's "linking of climate change and the Darfur crisis is simplistic" because causalities are complicated. He emphasized that while climate change does change livelihoods and cause disputes, proper institutions can settle things non-violently. De Waal considered that government mismanagement and militarization, not climate change, were main causes (2007(a)). The IPCC also questioned the characterization of Darfur as a climate war. In AR5, observations on climate-Darfur links were captured in box 12–5 of the chapter on "Human Security" (IPCC 2014, 755–91), with the conclusion that the conflict had multiple causes, and that the government of Sudan (GOS) played a key role. More recently, the online news portal *Breitbart* published a "fact check" of Joe Biden's speech, which described his claims about the climate-Darfur link, and labeled those as "false" and as an "alternative theory" devised by scientists (Moran 2020).

But what knowledge resources enable considering if Darfur was a climate war? Figure 5.1 shows the main actors, translations and resources involved.

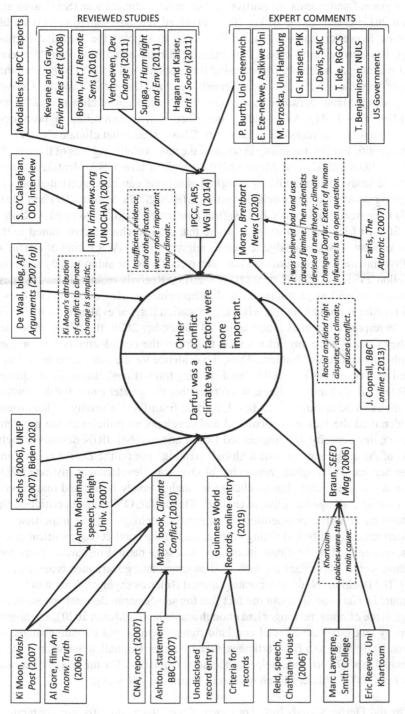

Figure 5.1 Main actors, translations and knowledge resources related to the general controversy about the climate-Darfur link

The general statements supportive of Darfur as a climate war (Srinivasan and Watson 2013; GWR 2019; Biden 2020) involved few and sometimes no visible resources. However, some cited Reid's speech to support the idea that climate change is a new security threat. For example, Braun noted how Reid "fingered global warming as a driving force behind the genocide in Darfur" (2006). Another resource was the CNA's *National Security and the Threat of Climate Change* (2007), which Mazo identified as the source of the concept of climate as a "threat multiplier" (2010, 81). Mjøs (2007) described how a committee of prominent American military officers – a reference to CNA – stated that climate changes are "a threat multiplier for instability in some of the most volatile regions of the world". Similarly, IRIN (2007) and Mazo (2010) described how CNA identified Darfur as a "case study of how existing marginal situations can be exacerbated beyond the tipping point by climate-related factors" (IRIN 2007). Mjøs (2007) and Mazo (2010) referenced Ki Moon's *Washington Post* article, and Mazo cited Al Gore's 2006 film *An Inconvenient Truth* to argue that climate change contributed to the Darfur conflict. The knowledge resources relevant to the GWR listing of Darfur as the "first climate war" are unclear, but each record must be submitted by someone and follow evidence requirements (GWR 2019). Records require at least witness statements, as well as photographic and video evidence. However, the GWR provided no information about the origins, publication date, or evidence for its Darfur entry. In response to email inquiries on 20 November 2020, the GWR responded that they cannot share any information to protect the confidentiality and privacy of applicants and record holders. Thus, the evidence for the entry remains hidden.

And what resources are used in the dissenting translations? Braun (2006) quoted Eric Reeves from Smith College, who stated that the greater cause for the conflict are policies in Khartoum, and Marc Lavergne from the University of Khartoum, who identified the lack of agricultural and development policies as the problem, observing that scarcity and neglect led Darfurians to rebel. IRIN described Sachs as one of the early ones to "put a global warming spin on the Darfur crisis", and quoted Sorcha O'Callaghan, researcher at Overseas Development Institute (ODI, UK), who lamented that climate change is "such a trendy issue" and that "everything is being packaged as climate change". On Darfur, O'Callaghan criticized the emphasis on resource competition as simplistic, highlighting the importance of considering complexity and listing other causes of the conflict (e.g. political grievances, race-related perceptions, inequality, access to natural resources, arms proliferation, youth militarization, lack of democracy and governance issues) (IRIN 2007). The Breitbart News inaccurately quoted Biden as saying: "What was Darfur all about? Darfur was all about the fact that the sub-Saharan desert, because of the change in the climate, no longer had enough arable land" (Moran 2020), and argued that many analysts blame racial and land rights disputes, not climate change, for the conflict, linking to a BBC article written by James Copnall, a Sudan analyst.[1] It then stated that *The Atlantic* helped popularize the idea of Darfur as a climate war. From Faris' 2007 article, Moran quoted in length the following:

Why did Darfur's lands fail? For much of the 1980s and '90s, environmental degradation in Darfur and other parts of the Sahel . . . was blamed on the inhabitants . . . Imprudent land use, it was argued, exposed more rock and sand, which absorb less sunlight than plants, instead reflecting it back toward space. This cooled the air near the surface, drawing clouds downward and reducing the chance of rain. "Africans were said to be doing it to themselves", says Isaac Held, a senior scientist at the National Oceanic and Atmospheric Administration.

(Moran 2020)

Moran also stipulated that "Faris wrote that climate scientists devised an alternative theory: climate change had changed Darfur's environment". To illustrate, he quoted Giannini's 2003 study on links between Indian Ocean temperature increase and Sahelian drought: "this was not caused by people cutting trees or overgrazing". On this basis, Moran concluded that the extent to which human activity can be blamed for environmental change remains an "open question".

So far, we have seen how the climate-Darfur link is debated in speeches and media contributions, and depends in particular on individual testimonies. A different modality emerged in 2014, when the IPCC synthesized the state of scientific research about the climate-Darfur link. The IPCC was established under the auspices of the World Meteorological Organization (WMO) to "collect, synthesize, and evaluate – knowledge about climate change" (Edwards 2013, 399). Many consider it "a model of global politics in which experts and expert knowledge, as politically neutral agents, were accorded significant power to define problems of global policy" (Miller 2004, 47) due to its legitimation through governments (Ibid., 58).

IPCC reports follow the *Procedures for the preparation, review, acceptance, adoption, approval and publication of IPCC reports* (IPCC 2013(d)), adopted at the IPCC plenary by government representatives. The workflow involves a first review of draft reports by experts, and a second review by governments and experts. All reviewers can comment, and the authors must respond (Edwards 2013, 399). The final report is considered by scientists and governments in plenary meetings. The IPCC's AR5 section on climate-Darfur links involved two coordinating researchers, six lead authors, four contributing authors, and two review editors. The authors were based in Australia, Chile, Colombia, Mexico, Norway, the Philippines, the UK and the United States.

What resources did the IPCC assemble to consider the climate-Darfur link? These are determined by the IPCC procedures, which identify what types of documents are considered, and outline the process of preparation, which involves a scoping of literature, a first draft, two rounds of comments by scientists and governments followed author responses, and finally the adoption of the final draft as part of the overall assessment report.

IPCC only reviews previous studies, and undertakes no data collection or methodological work. As inputs, the IPCC procedures prioritize "peer-reviewed scientific, technical and socio-economic literature" and recognize that reports by governments, industry, research institutions and international organizations, as well as conference proceedings, may be considered. Newspapers, blogs, social and broadcast media, and personal communications are explicitly excluded (IPCC 2013(d), 17).

In relation to climate-Darfur links, the IPCC reviewed Mazo's 2009 paper – identical with his 2010 book – as well as the studies by Kevane and Gray 2008; Brown 2010; Hagan and Kaiser 2011; Sunga 2011; Verhoeven 2011. The results were captured in box 12–5 (IPCC 2014, 773). The box described how Mazo identified climate change as a key cause, and how other studies disputed this by identifying multiple causes and by emphasizing the impossibility of isolating individual causes as most influential. The box also stated that most authors agreed that government practices matter more than climate, similar changes did not cause conflict in neighboring countries, and in the past Darfur had coped with climate variability without violence.

How does the IPCC arrive at these outcomes? Table 5.2 shows the visible steps to develop box 12–5: three drafts and two rounds of comments and responses in 2012–2014. The public drafts (IPCC 2012(a), 2013(a), 2013(c)) and compilations of comments (IPCC 2012(b), 2013(b)) allowed observing parts of the exchanges, and illustrated the process of translations and the evolution of the box. However, the preparation of the first draft is not visible. The final draft is very similar to the first, and thus a significant portion of the work undertaken remains invisible. Changes made in the final draft are not clearly traceable to the individual comments.

This illustrates the IPCC's general process for rejecting the attribution of the Darfur conflict to climate change. The consequence is that most of the climate-Darfur debate, which happens to a large extent in media and oral contributions, simply disappears. Even some reports by organizations (UNEP 2007; Bromwich et al. 2007) are not considered. The rejection is partly due to the fact that, due to its selection criteria for evidence, the IPCC reproduces the ways in which the five scientific publications sought to disassociate climate and the Darfur conflict. One might expect that, given IPCC's authority, this might have closed the debate. However, recent contributions (Adelphi 2015; Biden 2020) continued to cite Darfur as the first "climate war".

In the climate-Darfur debate, a specific sub-controversy developed on the assumptions of causality under which climate be considered a cause of conflict. Homer-Dixon, de Waal, Mazo and IPCC provided different accounts about the nature of causality in phenomena involving both ecological and social considerations.

In response to De Waal's identification of government mismanagement as the main cause (De Waal 2007(a)), Thomas Homer-Dixon argued in *Cause and Effect* that "additive models of causation are rarely valid in complex ecological-human systems". Instead, he described causation in such systems as "multiplicative" or "interactive", meaning that causes influence each other and are equally important. Furthermore, he stipulated that research on complex systems illustrates that small events can have large effects, and big events might have none, and that "disproportionality is the essence of nonlinear behavior". He also argued that in such contexts it is impossible to isolate causes or know the consequences of absence of causes. On this basis, Homer-Dixon considered claims about relative importance of causes of conflict invalid. He also criticized De Waal's use of "absent-factor-as-necessary-cause" when identifying government neglect as a key factor, arguing that something that did not happen cannot be considered a cause. He concluded that droughts

Table 5.2 IPCC process of drafts, comments and responses to develop box 12-5 on climate-Darfur links

First draft (IPCC 2012(a))	Comments (IPCC 2012(b))	Second draft (IPCC 2013(a))	Comments (IPCC 2013(b))	Final draft (IPCC 2013(c))/public version (2014)
Box 12-5. Climate and the Multiple Causes of Conflict in Darfur	J. Davis (SAIC): examples of impacts are detailed, while counter-examples general. Add discussion on when climate has not increased vulnerability to strengthen balance. Author response: there is little literature on cases where climate does not affect security.	As first draft	USA: change title of the box to "The multiple causes of the conflict in Darfur". Author response: box changed.	As second draft
		Box 12–5 demonstrates multiple resource and political dimensions of the Darfur conflict with no studies concluding that the conflict was climate-driven.	T. Ide: revise to: "that the conflict was mainly climate-driven". Author response: added the term "primarily".	Phrase deleted (reason unspecified)
Climate variability is popularly reported to be significant causes of the mass killing in the Darfur region that began in 2003 (see Mazo 2010): long term drought and vulnerability of the population to drought identified as the trigger and cause.		As first draft		Deletion of "long-term drought . . . identified as the trigger and cause"
Five detailed studies of the conflict conclude that climate variability and related environmental changes are proximate but not primary causes of the violence.		As first draft		Five detailed studies dispute the identification of the Darfur conflict as being primarily caused by climate change (Kevane and Gray 2008; Brown 2010; Hagan and Kaiser 2011; Sunga 2011; Verhoeven 2011).

(Continued)

Table 5.2 (Continued)

First draft (IPCC 2012(a))	Comments (IPCC 2012(b))	Second draft (IPCC 2013(a))	Comments (IPCC 2013(b))	Final draft (IPCC 2013(c))/public version (2014)
The detailed studies find that the violence in Darfur has multiple causes, including: • The legacy of past violence, which established groups that had a history of violent action, and a supply of weapons • Manipulation of ethnic divisions by elites in Khartoum • Weakening of traditional conflict resolution mechanism through government policies, and as a consequence of famines • Systematic exclusion of local groups from political processes, including of the Fur, Masalit, and Zaghawa ethnic groups • Limited economic development and inadequate provision of public services and social protection, stemming from governance and policy failures, political instability, and misuse of official development assistance • Desertification, declining productivity of arable land, and increased aridity (Brown 2010).	*P. Burth (Univ. of Greenwich): delete "including", because that implies that things are missing; add references. Authors: "including" changed to "notably"* *M. Brzoska (Univ. Hamburg): add extension of state power to Darfur in 1990, which contributed to tensions, escalation, and government alliance with militias. Integrate publications by Alex de Waal and Gérard Prunier. Author response: cannot add due to length constraints and because all causes and full chronology cannot be included.* *E. Ezenekwe (Nnamdi Azikiwe Univ.): Boko Haram is due to poverty, drought and desertification. Author response: cannot add cases due to page limits.*	As first draft, except references to all five studies added to the list of bullet points, and "including" replaced with "notably".	*T. Benjaminsen (NULS): desertification cannot be a conflict cause because Darfur greened before conflict, as shown by Kevane and Gray 2008 and others. Author response: reference to desertification is deleted due to dispute.*	"Desertification, declining productivity of arable land, and increased aridity" deleted; otherwise as second draft.

First draft	Comment / Author response	Second draft
All analyses agree that it is not possible to isolate any of these specific causes as being most influential (Hagan and Kaiser 2011; Kevane and Gray 2008; Sunga 2011; Verhoeven 2011). Most authors identify government practices as being far more influential drivers than climate variability, noting also that similar changes in climate did not stimulate conflicts of the same magnitude in neighboring regions, and that in the past people in Darfur were able to cope with climate variability in ways that avoided large scale violence.		As first draft
These studies therefore dispute the identification of the Darfur conflict as being caused by climate change, arguing that attributing this conflict to climate change masks the culpability of actors and the major drivers of insecurity.	*USA: replace "caused" with "aggravated"; attribution masks culpability, but most causes are climate-related. Author response: changed.*	Paragraph deleted
		As second draft

did contribute to the conflict in a "multicausal, interactive and nonlinear process", and that, based on evidence, explanations of the conflict must integrate climate as a causal factor (Homer-Dixon 2007).

De Waal responded by emphasizing that it is possible to estimate the relative importance of causes by comparing developments in different locations. He highlighted that across Sudan, areas with environmental degradation and rainfall decline often experienced less conflicts than areas with abundant natural resources. De Waal acknowledged that logically, neglect cannot be a cause of an event, but the failure to provide supplies qualifies as an action with consequences. Finally, he noted that "underlying" or "root" causes such as climate change are not always more significant than "immediate" and "brute" causes such as militarization (2007b).

Two years later, Mazo returned to questions of causality. He reviewed the debate between De Waal and Homer-Dixon. He described De Waal's first contribution as follows:

* The 1984–1985 famine in Darfur was caused by a drought, but no causal relation between climate and the drought has been proven.
* Drought, environmental degradation, low development and inefficient resource use led to food shortages. But Darfur's population doubled in 1985–2003, so the famine could not have happened due to resource scarcity.
* Instead, causes included lack of infrastructure, fertilizers and irrigation, and the food crisis turned into a famine due to government neglect.
* Violence was triggered by a deliberate policy, influx of weapons, Chadian militias, internal migration, economic reconfiguration, declining faith in government, loss of traditional authority and migration.
* Violence started when government mobilized militias to suppress resistance – drought and famine just made this easier. Thus, the government was the main culprit.

Mazo described this as a "textbook example of multiple and complex causation" (2010, 83). He then described Homer-Dixon's reply:

* In situations of complex causation, causes are multiplicative and/or interactive. "Mental manipulations" – hypothetically changing or subtracting one factor and seeing what happens – lead to bad results.
* Thus, there is no point to argue over relative importance of climate as cause of violence.
* De Waal implicitly acknowledges multi-causality and the impossibility of differentiating relative power of causes.
* Therefore, calling the government the main culprit is unsupportable.

Mazo then called De Waal's response a "reductionist argument" that did not address Homer-Dixon's point because he compared environmental conditions and violence within Darfur and saw empirical consistency between the militarization of government authority and conflict, but none between climate and conflict. Mazo commented that "this is true as far as it goes, but merely exemplifies . . . that what is meant by causation depends on why the question is asked". (Mazo 2010, 84). Mazo concluded that there is no disagreement that Darfur is an environmental conflict in some sense – just differences in perspective. He stated that if a cause is understood as both a necessary and sufficient condition, then Darfur cannot be seen as caused by climate change, but under less reductionist views of causality, climate change becomes a critical factor:

> To the extent that the question of climate change as a security issue is paramount, to say that other factors were equally, or even more, important politically or morally is not to deny that Darfur was a climate change conflict.
>
> (Mazo 2010, 85)

Climate change caused "the rains to fail" in Darfur

The translations forming the first controversy involved sweeping statements linking change and the conflict. But most controversies were focused on specific aspects of the climate-conflict links. The first specific aspect is that climate change caused rainfall decline and drought in the Sahel.

Faris (2007) and Ki Moon (2007) argued that climate change increased Indian Ocean temperatures, causing rainfall decline and drought in the Sahel. This was also raised by UNEP's 2007 report, which introduced the data in Table 5.3 and described how Darfur had a "irregular but marked rainfall decline" (UNEP 2007, 84), amounting to evidence for climate-induced lack of rain (Ibid., 9).

UNEP compared rainfall for two 30-year periods, 1946–1975 and 1976–2005, recorded at three stations, noting reductions ranging from 16 to 34 per cent (Ibid., 60). The report also noted that since records began, the ten-year moving average rainfall for El Fasher declined from 300 mm/year to 200 mm/year, the last time rainfall exceeded 400 mm/year was in 1953, and that a decline of 30 per cent over 80 years was recorded in northern Darfur (Ibid.).

However, De Waal (2007(a)) and Kevane and Gray (2008) questioned whether rains declined. De Waal argued that drought is a permanent feature of Darfur's

Table 5.3 Long-term rainfall reduction in Darfur

Rain gauge location	Average annual rainfall (mm) 1946–1975	Average annual rainfall (mm) 1976–2005	Reduction	Percentage
El Fasher, Northern Darfur	272.36	178.90	– 93.46	– 34%
Nyala, Southern Darfur	448.71	376.50	– 72.21	– 16%
El Geneina, Western Darfur	564.20	427.70	– 136.50	– 24%

Source: UNEP 2007, 60

climate, and that the region has successfully coped with drying in the past. Kevane and Gray tested whether: (1) rainfall levels in Darfur reduced before the conflict; and (2) whether there is any correlation between rainfall change and conflict. On both counts, the authors reached negative conclusions about lack of rain prior to the conflict.

Figure 5.2 summarizes the actors, translations and resources involved in debating whether rains failed prior to the conflict. Faris (2007) and Ki Moon (2007) referred to a study on climate-drought links in Africa led by Alessandra Giannini of Columbia University and published in *Science* (Giannini et al. 2003). The study did not consider climate-Darfur links specifically, but considered sea surface temperature data and an atmospheric general circulation model developed by National Aeronautics and Space Administration (NASA's) Goddard Space Flight Center to illustrate that change of Indian Ocean temperature disrupts African monsoons, driving rainfall decline and Sahelian droughts (Ibid., 1027). Giannini et al. was the first reference to a scientific study in the climate-Darfur debate. Faris cited statements by Giannini to present the drying of Darfur as climate-driven. Similarly, Ki Moon reflected the main findings of Giannini et al. as follows:

> Subsequent investigation found that [precipitation decline] coincided with a rise in temperatures of the Indian Ocean, disrupting seasonal monsoons. This suggests that the drying of sub-Saharan Africa derives, to some degree, from man-made global warming.
>
> (Ki Moon 2007)

To support the rainfall decline component, UNEP 2007 used precipitation data to demonstrate rainfall decreases prior to the conflict (see Table 5.3). "Data" are numerical representations of information in ways that make involve modularity, automation, variability, transcoding, manipulability and mutability (Bates et al. 2016). They are physical magnetic atoms that can be transmitted via electromagnetic signals (Ibid.). They are a "product of a particular set of practices" (Ibid., 3), and have "dimensionality, weight, and texture" (Edwards 2013, 84) and material consequences (Bates et al. 2016, 3). Law and Mol describe data as "mutable mobiles", which can move between sites, and which practitioners can mix and adapt for different purposes (Law and Mol 2001). UNEP 2007 was the first contribution to involve rainfall data, to identify its sources and to present it in a table. This way, rainfall data becomes a vehicle for representing Darfur. Others regularly referred to data as an unspecified entity. For instance, Ki Moon (2007) cited unspecified "UN statistics" that indicate precipitation reduction in Darfur.

But how were these data generated, processed and integrated into the translations? UNEP data was collected by rainfall gauges at three measurement stations in Darfur. These were, at the time, the only continuously monitored stations in an area of 800,000 km^2 (UNEP 2007, 60) – the size of Turkey. Information on the stations can be found in WMO's *Volume A – Observing Stations*, which lists the world's weather stations (WMO 2014(a)), and in WMO's online database OSCAR (Observing Systems Capability Analysis and Review Tool).[2] These indicate, for

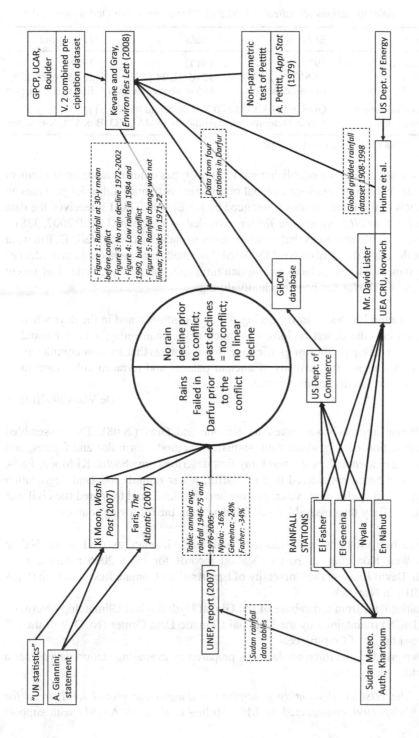

Figure 5.2 Actors, translations and knowledge resources related to the controversy about rainfall decline prior to the Darfur conflict

Table 5.4 Rainfall stations identified by UNEP (2007) as sources for Darfur rainfall data

Station	El Fasher	Nyala	El Geneina
Established	1921	1940	1937
Location	13.617°N, 25.333°E	12.05°N, 24.88°E	13.48°N, 22.45°E
	El Fasher airport	805 m above sea level	El Geneina airport
Observation networks	GOS, GSN, CLIMAT(C), RBCN and RBSN(ST)		
Station operators	Sudan Meteorological Authority (SMA), P.O. Box 574, Khartoum		

Sources: WMO 2014(a), 2014(b), *OSCAR*

each station, the date of establishment, location, types of instruments, measurement intervals and the linkages with global observation networks. Table 5.4 provides an overview of the three stations referenced by UNEP 2007. UNEP received the data as *Sudan Rainfall Station Data Tables*, provided by the SMA (UNEP 2007, 338).

These are the resources that enabled actors to claim that "rains failed". But what about the critical contributions? De Waal described that "archeological evidence" from rock paintings, palaces and agricultural ruins show how, instead of recent desiccation, Darfur has been gradually drying:

The region has been slowly drying out over centuries, and in the days when travel into the desert regions was much easier than today, a visitor could observe rock paintings of giraffes and rhinos in areas that are now completely barren, and see the remains of ancient palaces and terraced cultivation in areas now too dry to support life.

(De Waal 2007(a))

A different approach was taken by Kevane and Gray (2008). They assembled rainfall station data, gridded data, statistical methods, formulas and figures, and compared evidence. Kevane and Gray first described how Sachs, Ki Moon, Faris, UNEP and Stern considered that in Darfur, lower rainfall and land degradation intensified struggles over water, pasture and farmland, and triggered the civil war (2008, 1). They then assembled rainfall data from the following sources:

- Data from stations in El Fasher, Nyala and Geneina, as well as En Nahud in West Kordofan, as part of station records for 1881–2006 received from Mr. David Lister at the University of East Anglia's Climate Research Unit (UEA CRU) in Norwich.
- Station data from a database of the Global Hydrological Climatology Network (GHCN) maintained by the National Climatic Data Center (NCDC) of the US Department of Commerce.
- Two gridded precipitation datasets, prepared by averaging station data across a grid:

  - The *Historical monthly precipitation dataset for global land areas for 1900–1998* constructed by Mike Hulme et al. at UEA CRU with support of the UK Department of Environment, Transport and the Regions. Hulme et al. used UEA CRU and US Department of Energy data encompassing 11,880 stations. They used *Thiessen polygon weights*, a method for determining mean area precipitation, to average station data in grid boxes sized 2.5° latitude times 3.75° longitude (Hulme 1999). Kevane and Gray accessed this dataset via an FTP-based process described by Hulme 1999, and identified four observation points for Darfur located at crossing points of 12.5° and 15° latitudes and 22.5° and 26.25° longitudes, which they referred to as the "Hulme nodes".

- *Version 2 Combined Precipitation Data Set* of the Global Precipitation Climatology Project (GPCP), which blends estimates from satellites and rain gauges organized into a 2.5° x 2.5° global grid. From it, to complement the Hulme nodes, the authors identified four "GPCP nodes" across Darfur.

Kevane and Gray also compared their data with other studies, mainly, the 1991 book *Drought and Famine Relationships in Sudan: Policy Implications* by Teklu and Von Braun at International Food Policy Research Institute (IFPRI), which indicated rainfall patterns similar to the UEA CRU data, and information on historical rainfall patterns from a 2001 article *Climatic and environmental change in Africa during the last two centuries* by Sarah Nicholson published in the Journal *Climate Research*.

While the presentation of rainfall data in UNEP 2007 was limited to one table, Kevane and Gray presented several different visual presentations of rainfall data. Their figure 2 is a line chart presenting annual rainfall in mm at the four Darfur stations between 1972 and 2002, and identifies mean rainfall with a dashed line. Kevane and Gray's Figure 3 provided a similar line chart presentation for the GPCP nodes for 1979–2002. Based on these, the authors made three observations: rainfall was close to a 30-year mean before conflict outbreak, there was no decline in 1972–2002, and though rainfall was low in 1984 and 1990, no conflicts occurred (Ibid., 5). The two figures reinforce each other by illustrating the same trend based on different sources. This is the first part of the refutation of the argument that the conflict was preceded by lack of rainfall.

Kevane and Gray further emphasized the lack of correlation between low rainfall and conflicts (their Figure 4), which compared monthly rainfall for 1984, 1990 and 2002 in the GPCP nodes, illustrating how, when rainfall was much lower in 1984 and 1990, no conflicts occurred. A further presentation was their Table 1, which showed rainfall trends for 22 different 50-year periods at the four stations. All except En Nahud showed declines when starting in the 1940s and 1950s – years of high rainfall. However, starting in early 60s, no stations demonstrate "statistically significant" trends, established by the authors as P value of 0.05 or lower (Ibid. 6). They considered this as further evidence against declining rainfall.

Kevane and Gray's Figure 5 is a set of four line charts, each showing the average rainfall in mm per year in one of the four Darfur stations between 1940 and 2002. The authors used dashed lines to show the mean rainfall for 1940–1972 and 1971–2002, indicating a sudden shift from a higher average in 1940–1972 to a lower one in 1971–2002. On this basis, the authors argued that assumptions about a gradual rainfall decline are misleading and that there was a "structural break" in rainfall in the 1970s, when it shifted from a higher to a lower long-term mean.

Finally, Kevane and Gray's table 2 showed when the rainfall shifted in station data and at the Hulme nodes. To estimate the timing of the rainfall shift, they applied the *non-parametric test of Pettitt* – a statistical test that detects the timing of sudden changes in mean values of variables. For each station and node, the table identified the years of the shifts (between 1964 to 1971), and mean rainfall for 1940–1971 and 1972–2002. The authors then compared this finding with a study by Mahe et al. (2001), which found a similar shift in rainfall in West Africa around 1970.

These are the modalities through which Kevane and Gray questioned the notion that rains failed. Rainfall data was collected at four stations in Darfur and by satellites. The data was transmitted to and processed by scientific institutions and government agencies, mostly in the United States or Western Europe. The processing involved changing data into different forms – including conversion into "nodes", online publication, and translation into figures with statistical tools such as regression analysis, non-parametric test of Pettitt, and crosstabs. On this basis, the authors argued that rains did not fail prior to the conflict, that when they failed in 1984 and 1990 no conflicts occurred and that the impression of a rainfall decline results from the structural break in the 1970s. Kevane and Gray's contribution was the first peer-reviewed article addressing the climate-Darfur link.

However, not everyone accepted Kevane and Gray's results. In *Climate Conflict* (Mazo 2010), Mazo countered the criticisms and rearticulated the climate-Darfur mechanism by synthesizing existing materials and comparing evidence. He introduced a new dimension to the climate-Darfur link: while the links between rainfall and conflict might not be clear, there is a correlation between temperature and conflict across Africa, and thus climate change can be considered a key cause.

Against Kevane and Gray's criticism of the rainfall component, in particular the finding that rainfall did not decline prior to the conflict, Mazo argued that even short-term fluctuations can "contribute to environmental stress and conflict". For this, he referred to the IPCC Working Group II's fourth assessment report (IPCC 2007, 299), which observed that multi-year oscillations of rainfall have become more severe and frequent since the 1980s. On this basis, Mazo argued that linear long-term rainfall decline is not needed to cause serious impacts. Second, against Kevane and Gray's observation that in 1980–1984 and in 1990, droughts did not provoke conflicts, Mazo referenced UNEP's 2007 observation that 15 of 29 local conflicts over grazing and water rights occurred in 1980–1984 and 1990–1991, indicating that water scarcity-related conflicts did erupt during droughts.

Third, against Kevane and Gray's finding that there was no correlation between rainfall breaks and conflicts across Africa, Mazo described Burke et al. (2009), a study published in *Proceedings of the National Academy of Sciences* (PNAS), which stipulated a correlation between temperature and civil war in Sub-Saharan Africa in 1980–2002. He highlighting that the correlation held true "when controlled for a range of data sets, models and variables, including precipitation, per capita income and degree of democracy". Mazo also quoted a blog in which Kevane acknowledged that they did not have enough data to consider temperature and noted the differences between temperature and precipitation change trends (Kevane 2009).

With this, Mazo sought to do three things: (1) Argue that even short-term rainfall fluctuations can trigger conflicts (while Kevane and Gray focused on long-term trends); (2) Replace the rainfall component of the climate-Darfur link with temperature (if the rainfall-conflict link is controversial, then Burke et al. have demonstrated a temperature-conflict correlation); (3) Question Kevane and Gray's statement that no conflicts broke out in 1984 and 1990 by presenting UNEP's observation that most water-related clashes in Darfur occurred during those times.

This approach is new in the climate-Darfur debate. Mazo first reintroduced advocates of the climate-Darfur link, and then mobilized IPCC 2007; Burke et al. 2009; UNEP 2007 against Kevane and Gray. Mazo did not, as Kevane and Gray, question the conditions of production of previous statements, or mobilize new data or methods. Rather, he introduced findings by others that contradicted Kevane and Gray, and quoted Kevane himself on the gaps in their study.

Resource scarcity was a key factor in the conflict

Another component of the climate-Darfur mechanism is the notion that climate change increased resource scarcity, in particular of water and arable land. For example, Klare argued that "global climate change and dwindling natural resources are combining to increase the likelihood of violent conflict over land, water and energy" (Klare 2006). This logic was applied to Darfur. For example:

> Long periods of drought in the 1970s and 1980s in Sudan's Northern Darfur State . . . resulted in deep, widespread poverty and, along with many other factors such as a breakdown in methods of coping with drought, has been identified by some studies as a contributor to the current crisis there.
>
> (Stern 2006, 112)

Similar assumptions were expressed by John Reid, whom three other contributions (Russell and Morris, Klare and Baldauf) quoted:

> The blunt truth is that the lack of water and agricultural land is a significant contributory factor to the tragic conflict we see unfolding in Darfur.
>
> (John Reid in Russell and Morris 2006)

Along the same lines, Mr. Ki Moon considered that "it is no accident that the violence in Darfur erupted during the drought" (2007), and Baldauf argued that while the conflict is often presented in ethnic or religious terms, the competition for resources is equally important (2006). Similarly, Professor Mjøs describes how:

> the UN Secretary-General, Ban Ki-Moon, said in his careful way that "when resources are scarce – whether energy, water or arable land – our fragile ecosystems become strained, as do the coping mechanisms of groups and individuals. This can lead to a breakdown of established codes of conduct, and even outright conflict".
>
> (Mjøs 2007)

Ambassador Mohamad of Sudan also evoked resource scarcity:

> The major cause of the question of Darfur is the environmental degradation from climate change . . . Darfur is a classic case of climate change. People have witnessed gradual degradation of the environment and erosion of the resources and desertification and drought that was going on for a long time, since the beginnings of the seventies.
>
> (Straw 2007)

The resource scarcity thesis was articulated more specifically by UNEP, which saw "complex but clear" linkages between environment and the conflict, and identified climate change, land degradation, and resource competition as root causes (UNEP 2007, 20). The report also noted "strong indications" that desertification is one underlying cause of war (Ibid., 58) and concluded:

> The scale of historical climate change as recorded in Northern Darfur is almost unprecedented: the reduction in rainfall has turned millions of hectares of already marginal semi-desert grazing land into desert. The impact of climate change is considered to be directly related to the conflict in the region, as desertification has added significantly to the stress on the livelihoods of pastoralist societies, forcing them to move south to find pasture.
>
> (Ibid., 60)

These assumptions were questioned by Brown (2010). Based on satellite data and the NDVI, He concluded that Darfur did not suffer from resource scarcity and that vegetation levels improved prior to the conflict, indicating higher agricultural production.

But what knowledge resources are involved in this part of the mechanism? Figure 5.3 shows the actors, translations and resources relevant to the resource scarcity controversy. Some cited the report *An Abrupt Climate Change Scenario and Its Implications for US National Security*, commissioned by the DoD and prepared by consultants Peter Schwartz and Doug Randall of the Berkeley-based Global

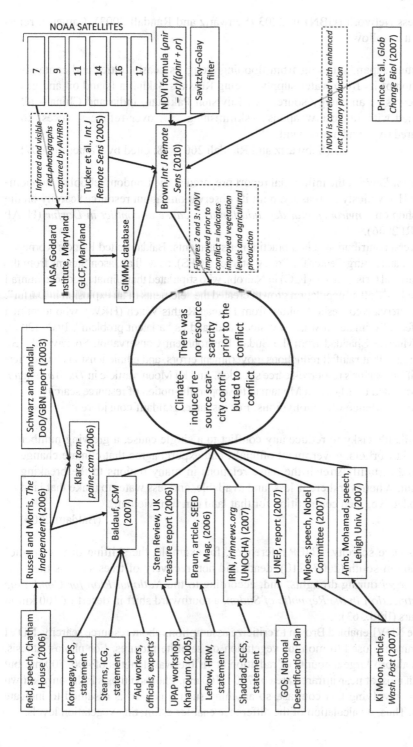

Figure 5.3 Actors, translations and knowledge resources related to the controversy about resource scarcity prior to the Darfur conflict

Business Network (GBN) in 2003 (Schwartz and Randall 2003). Cameron references it as follows:

> Picture Japan, suffering from flooding along its coastal cities and contamination of its fresh water supply, eyeing Russia's Sakhalin Island oil and gas reserves as an energy source . . . Envision Pakistan, India and China – all armed with nuclear weapons – skirmishing . . . over refugees, access to shared river and arable land.
>
>   (Schwartz and Randall 2003, 18; cited by Cameron 2007)

The *Stern Review*, the influential report prepared at the London School of Economics for Her Majesty's Treasury, built its observations about resource scarcity on the workshop on *Environmental degradation as a cause of conflict in Darfur* (UPAP and PRI 2006).

Several contributions cited practitioner statements. Baldauf cited Francis Kornegay from Johannesburg Center for Policy Studies (JCPS), as well as Jason Stearns from the International Crisis Group (ICG) in Nairobi, who stipulated that climate, reduced rainfall and land, as well as population growth created the "elements of an explosion in Darfur". Braun interviewed Leslie Lefkow from Human Rights Watch (HRW), who identified the effect of climate on water, livestock and land as "a latent problem". IRIN (2007) cited Muawia Shaddad from the Sudan Environment Conservation Society (SECS), who argued that rainfall reductions, government errors, and ethnic tensions found "fertile soil" in Darfur's resource-scarce situation. And Ki Moon's article in *The Washington Post* was cited by Mjøs and Mohamad to support the notion of resource scarcity.

Some evidence was anonymous. For instance, Baldauf considered:

> While it's risky to reduce any conflict to a single cause, a growing number of aid workers, government officials, and experts agree that climate change could certainly stretch the tense relations in many regions to the breaking point. Whenever there is less land available, and less water to make that land productive, then competition for that land can turn violent.
>
>   (Baldauf 2006)

On resource scarcity, UNEP referenced, for example, the shifting of tree species 50–200 km south (2007, 210), death and non-recovery of trees such as the *Acacia Senegal* during droughts, and, according to the *National Plan for Combating Desertification in the Republic of Sudan*, a southward shift in desert of 100 km in 40 years (Ibid., 63).

The NDVI enabled Brown to criticize assumptions about resource scarcity. NDVI is commonly used to monitor vegetation on satellite pictures. For photosynthesis, plants absorb large amounts of radiation at visible wavelengths of 400–700 nm but re-emit most at near-infrared area of 700–1100 nm. Photosynthesis is the only known process involving this contrast, so anything such a pattern is assumed to indicate plants. Thus, a calculation of the difference enables estimating vegetation levels.

Brown used 200 infra- and visible-red photographic images produced at NASA Goddard Space Flight Center in Greenbelt, Maryland, and the Global Land Cover Facility (GLCF) of the University of Maryland. The images were part of the Global Inventory Modeling and Mapping Studies (GIMMS) dataset, built on data from instruments known as the Advanced Very High Resolution Radiometer (AVHRR) on NOAA polar-orbiting satellites 7, 9, 11, 14, 16 and 17. The AVHRRs capture images at 4 km resolution. The images were processed at the Goddard's Laboratory of Terrestrial Physics, Biospheric Sciences Branch, where a team built a dataset for July 1981 to December 2004 for all continents, calibrated and corrected it for orbital drift, viewing geometry and volcanic aerosols, and converted it into an 8 km resolution. The results of this were presented in 2005 in the *International Journal of Remote Sensing*.

Based on the images, Brown calculated the NDVI for Darfur using the formula $(\rho_{nir} - \rho_r)/(\rho_{nir} + \rho_r)$ and produced two figures (Figures 5.2 and 5.3). Figure 5.2 is a line chart on "spatially averaged NDVI values" in Darfur between 1982 and 2006. For this, Brown applied a Savitzy-Golay filter, a tool to smooth digital data points. The figure shows that NDVI improved in Western Darfur and remained steady in the north. Figure 5.3 is a map that shows NDVI for the growing season in July-October for 1984, 1992, 2001 and 2002. Its intention was to illustrate that vegetation coverage was at its highest in years before the conflict. Brown also compared the GIMMS data with two other vegetation index datasets: the 1 km SPOT VEG-ETATION, and the 30 m Landsat Enhanced Thematic Mapper Plus, describing the three as "well correlated" (Ibid., 2516). Finally, Brown cited a 2007 study on Sahel desertification, published in *Global Change Biology* by Prince et al., which identified a positive correlation between NDVI and agricultural production. On that basis, Brown concluded that Darfur did not face significant resource scarcities prior to the conflict (ibid., 2517).

Resource scarcity drove farmers and pastoralists into fighting

The previous sections described how the climate-Darfur link was built on the notions that rains failed and resources were scarce. However, since wars involve people, the link required a human component to connect climate, rainfall and resource scarcity with the conflict. Thus, it was stipulated that resource scarcity impacted the balance between farmers and pastoralists, triggering migrations and the conflicts.

In 2006, Sachs described Darfur as an arid zone with populations of rain-dependent pastoralists and farmers, and Baldauf (2006) considered that resource problems pitted them against each other. Others characterized climate and farmer-pastoralist relations in these terms:

> We are already seeing this: a contributing factor to the conflict in Darfur has been a change in rainfall that pitted nomadic herders against settled farmers.
> (Cameron 2007)

Increasing drought cycles and the Sahara's southward expansion have created conflicts between nomadic and sedentary groups over shortages of water
and land.

(Braun 2006)

Similarly, UNEP highlighted that rainfall decline stressed pastoralist societies,
forcing them to migrate, and thus contributed to conflict (2007, 9, 58), and UNEP's
findings were cited by others, including Mamdani (2009, 207).

As the previous two components, the farmer-pastoralist component also
turned controversial. IRIN (2007) considered pastoralist-farmer assumptions simplistic, and Selby and Hoffmann criticized them as "stereotypical",
"outdated" and "traditionalist", describing Sudanese livelihoods as "hybrid,
dynamic, globally integrated . . . and distinctly modern" with revenues from
agricultural exports, remittances, development aid, public employment and oil
production (2014, 365).

Figure 5.4 provides an overview of the actors, translations and knowledge
resources within this controversy. Most statements identifying farmer-pastoralist
tensions cited no resources, and only two resources were visible. First, Baldauf
quoted an "anonymous North African diplomat" describing Darfur as follows:

On one side you have herders. On the other side you have farmers. And with the
spread of weapons in the region, it becomes very dangerous and hard to control.

(Baldauf 2006)

The other resource was a description of a 1985 meeting between Alexander De
Waal and Sheikh Hilal Abdalla in Darfur. Faris describes how De Waal told him
"a story that, he says, keeps coming back to him" (2007). De Waal had met a "bed-
ridden and nearly blind" Sheikh during his PhD fieldwork. The Sheikh spoke about
sand ruining lands, and rains washing away soils. Before the drought, farmers and
pastoralists shared land, but once droughts ruined the farmlands, farmers prevented
pastoralists from grazing, triggering tensions and war. Faris told that the Janjaweed
attacks were led by Musa Halil, sheikh Abdalla's son, whose aggression originated
in his father's fears about climate change.

In the critical contributions, IRIN described how UNEP acknowledged that
Sudanese livelihoods are more complex than pastoralist-agriculturalist relations,
and cited Geoffrey Dabelko, who stated that while pastoralist-agriculturalist competition is key to many African conflicts, it would be important to avoid simplistic
and deterministic approaches (Ibid.). In their criticisms, Selby and Hoffmann referenced, in particular (2014, 365):

- De Waal's 1989 book *Famine That Kills* to argue that "the subsistence tribal
  peasant no longer exists in Darfur, and has not done for decades";
- The book *Darfur: Livelihoods Under Siege* by Helen Young et al., to highlight
  that the nomads-farmer distinction is untenable because most farmers raise livestock, and most pastoralists farm;

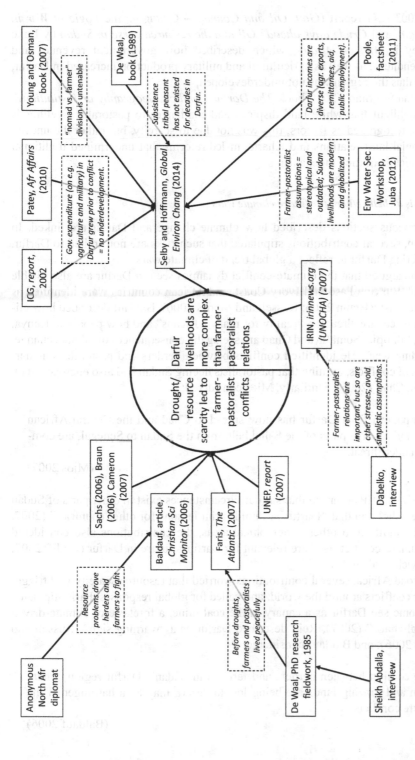

Figure 5.4 Main actors, translations and knowledge resources related to the controversy about farmer-pastoralist relations in context of the Darfur conflict

- A 2002 ICG report (*God, Oil and Country – Changing the Logic of War in Sudan*), and *Crude days ahead? Oil and the resource curse in Sudan* by Luke Patey in *African Affairs*, which described how government revenues and expenditure, as well as agricultural and military production increased in Darfur, and thus the region was not underdeveloped;
- Osman Suliman's 2010 book *The Darfur Conflict: Geography or Institutions?* to highlight that while land disputes and challenges to pastoralist livelihoods were recognized as factors, this was not due to scarcity, but rather the uneven colonial land relations and Khartoum-led reforms that undermined traditional conflict resolution mechanisms.

Darfur is a "bellwether" for future climate wars

The previous sections described how climate change and Darfur are linked. In addition, several contributions stipulated that such risks are not limited to Darfur, considering Darfur to reflect a global trend of climate wars.

Some argued that the climate-conflict dynamics seen in Darfur are also visible in Chad, Burkina Faso and Ivory Coast, and African countries were identified as at highest risk (Braun 2006; Russell and Morris 2006). Baldauf described how climate impacts are already a "reality for many Africans" and how people in Kenya, Sudan, Ethiopia, Somalia and Chad are seeing the consequences of climate change, "including war". He identified conflicts between herders and pastoralists as harbingers of conflicts, arguing that pastoralists fleeing Sudan had also encroached on farms in Chad (2006). Similarly, Mjøs considered:

> The pattern from Darfur has now spread to Chad and the Central African Republic. Large parts of the Sahel belt, from the Sudan to Senegal, are coming under threat.
>
> (Mjøs 2007)

Similarly, UNEP observed that similar circumstances exist in other parts of Sudan and the Sahel and that "Darfur . . . holds grim lessons for other countries" (2007, 329). In addition to other geographical areas, the contributions also considered how climate-conflict risks are relevant for Darfur's future in Darfur (UNEP 2007; Bromwich et al. 2007).

Beyond Africa, several contributions worried that resource scarcity will trigger similar conflicts around the world, and called for global responses. Faris stipulated that "some see Darfur as a canary in the coal mine, a foretaste of climate-driven political chaos" (2007), Reid described Darfur as a "warning sign" (Russell and Morris 2006), and Baldauf considered:

> The conflict between herders and farmers in Sudan's Darfur region, where farm and grazing lands are being lost to desert, may be a harbinger of the future conflicts.
>
> (Baldauf 2006)

At Major Economies Meeting, President Sarkozy presented the following picture:

> In Darfur, we see this explosive mixture from the impact of climate change, which prompts emigration by increasingly impoverished people, which then has consequences in war . . . If we keep going down this path, climate change will encourage the immigration of people with nothing [to] areas where the populations do have something, and the Darfur crisis will be only one crisis among dozens of others.
>
> (Nicolas Sarkozy, quoted Ingham and Hood 2008)

As preparation for global climate-security risks, Reid suggested that British armed forces prepare for resource conflicts, humanitarian work, peacekeeping and new conflicts (Russell and Morris 2006). David Cameron emphasized prevention, arguing that "politicians have a duty to prepare for [the] consequences [of climate change] in terms of domestic and international security", called for UK climate targets, and criticized Labour for dropping climate commitments. He also called for investments in renewable energy to reduce supply vulnerability, focus on "preventing and addressing climate change in the developing world", and assessments of climate conflicts in defense planning (Cameron 2007). Finally, Mjøs considered that "climate and the environment have . . . become one of the threats "to international peace and security", suggesting that the UNSC should respond (Mjøs 2007). In contrast, Klare stated that military buildup has limits, and urged cooperation to reduce emissions (Klare 2006).

Few contributions criticized the aforementioned arguments directly, but Kevane and Gray also tested assumptions about climate-related conflicts in other parts of Africa. They observed that while 22 of 28 countries had rainfall breaks, there was no obvious relationship between those and conflicts (2008, 8).

The actors, key translations and knowledge resources within this controversy are illustrated in Figure 5.5. To link war in Darfur and other conflicts in the region, Baldauf again quoted the anonymous North African diplomat as follows: "the fighting in Chad, and the fighting in Darfur are the same" (2006). Ki Moon, citing Sachs, connected Darfur and other conflicts by arguing that

> the stakes go well beyond Darfur. Jeffrey Sachs, the Columbia University economist and one of my senior advisers, notes that the violence in Somalia grows from a similarly volatile mix of food and water insecurity. So do the troubles in Ivory Coast and Burkina Faso.
>
> (Ki Moon 2007)

Sachs himself considered that, based on "several studies", the problem is not unique to Darfur because rainfall decline in Sub-Saharan Africa is associated with rise in conflict likelihood (Sachs 2006). Sachs was possibly referring to *Economic Shocks and Civil Conflict: An Instrumental Variables Approach* by Miguel et al. (2004), which investigated the correlation between rainfall, economic impacts and conflict in Sub-Saharan Africa. Descriptions of Africa as the region at most risk

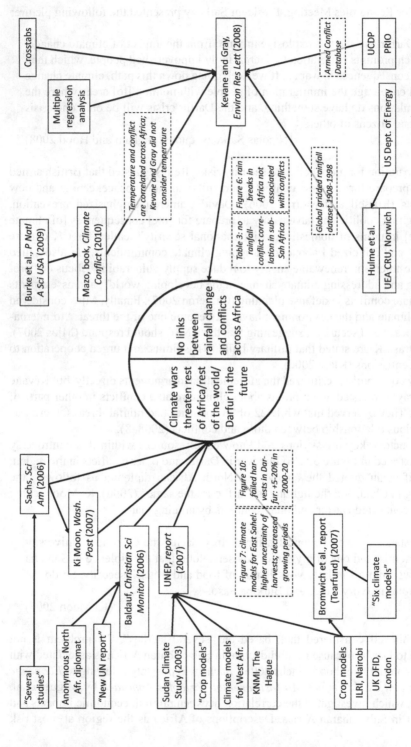

Figure 5.5 Main actors, translations and knowledge resources related to the controversy about whether Darfur is a warning sign for future climate wars

from climate-related conflicts also referenced a "new UN report" attributed to the UNFCCC secretariat (its origins remains unclear), which stated that in Africa climate change might destroy five per cent of crops, 25–40 per cent of natural habitats, and 30 per cent of coastal infrastructure (Baldauf 2006).

UNEP cited model simulations which predict temperature increases, rainfall decreases, and crop yields changes. They described a 2003 *Sudan Climate Study*, which modelled changes in temperature and rainfall for Kordofan for 2030 and 2050, projecting a 0.5–1.5°C increase in annual temperature and a five per cent rainfall reduction. These findings were merged with crop models to estimate changes in agricultural production, projecting a "potentially disastrous decline" of 70 per cent. UNEP also highlighted that other models predicted similar problems for all of Africa, though with different rainfall predictions (UNEP 2007, 61). These were: a model on West Africa run by the Royal Netherlands Meteorological Institute (KNMI), a *Simulation of Sahel Drought in the 20th and 21st Century* published in 2001 by Isaac Held et al., and a *Mapping Climate Vulnerability and Poverty in Africa*, a report by Thornton et al. in 2007 for the International Livestock Research Institute (ILRI) and the UK Department for International Development (DFID). Thornton et al. focused on changes in the growing season and predicted that failed harvests in Darfur would increase by 5–20 per cent between 2000 and 2020 (UNEP 2007, 61). On this basis, UNEP outlined the following forecast for the Sahel:

> Historical data, anecdotal field reports and modelling all point to the same general trend. Overall, rainfall is becoming increasingly scarce and/or unreliable in Sudan's Sahel belt, and this trend is likely to continue. On this basis alone, large tracts of the Sahel will be severely impacted by declining food productivity over the next generation and beyond.
>
> (UNEP 2007, 61)

Similarly, a report by the UK-based non-governmental organization (NGO) Tearfund (Bromwich et al. 2007) integrated information from KNMI models, captured in the report *Changes in extreme weather in Africa under global warming* (47). Figure 7 of the Tearfund report (see Ibid., 45) compared observed precipitation changes with projections of six climate models between 1910 and 2100.[3] The report concluded that while the models differed in their predictions for the Sahel, they predicted concurring increases in variability and frequency of failed harvests and decreased growing periods, and some showed significant downward trends (Ibid.)

Tearfund also referred to Thornton et al. to stipulate:

> Climate models predict that in Darfur the length of growing periods will reduce and . . . failed harvests will increase (e.g. a 5–20 per cent reduction in growing period from South to North Darfur by 2020 compared to 2000 levels . . . and >20 per cent for central Darfur by 2050).
>
> (Ibid., 69)

Bromwich et al. also included a map of Africa (their figure 10), replicated from Thornton et al., to visualize the projected changes in growing period between 2000 and 2020 due to climate change. The map showed losses of over 20 per cent across the Sahel and southern Africa. The report then connected the projections with risks of violence:

> Climate change models indicate that the climatic variability will increase and length of growing periods will decrease. These impacts will cause a reduction in yields and an increase in crop failures – acting as triggers for further violence.
>
> (Ibid. 26)

In their criticisms, Kevane and Gray used conflict data to consider whether precipitation change is correlated with conflicts. The authors used a dataset from Miguel et al. (2004), who, in turn, drew on the Armed Conflict Dataset of the Uppsala Conflict Data Program (UCDP) and Peace Research Institute Oslo (PRIO), as well as the Collier War Measure. Kevane and Gray tested whether rainfall breaks were associated with conflicts in Burkina Faso, Chad, Ethiopia, Mali, Niger, or Sudan. They introduced a line chart showing the average rainfall for those countries in 1940–1998, based on simple average precipitation at the Hulme nodes in each country (see Kevane and Gray 2008, 8). Based on this, the authors stated that all countries had a break in rainfall, but Mali, Burkina Faso, and Niger were relatively peaceful, and thus the breaks were not associated with conflict. Kevane and Gray also tested whether rainfall breaks and conflict are correlated across sub-Saharan Africa. They used averages of all Hulme nodes for 38 African countries, and calculated the number of conflict years based on Miguel et al. (2004) conflict data. They captured the results in a table, concluding that "crosstabs and multiple regression analysis of various measures of war against the presence or absence of a structural break suggest no pattern" (Kevane and Gray 2008, 8).

The Sudanese government bears less responsibility for the conflict

So far, we have seen how the contributions have framed Darfur as a climate war, articulated connections between climate and the conflict, and described Darfur as a "warning sign" for future climate wars. The last controversy is about the implications of climate-conflict links on the responsibility of the Sudanese government. Some argued that the government was helpless to prevent this war, and that others might bear at least some responsibility. Faris considered that high emitting countries might bear some responsibility (2007). Similarly, Sudanese Ambassador Osman claimed at the UNSC that climate-induced drought and desertification caused the conflict as they forced herdsmen to encroach on farmland, that some conflict parties exploited the situation for political causes, and that the conflict could have been prevent by spending the billions used on peacekeeping In Darfur on reducing the effects of desertification and drought (UNSC 2011, 34).

Others argued that such assumptions let Khartoum "off the hook" by enabling the government to blame climate change and high-emitting countries. The critics emphasized the importance of not whitewashing the government (IRIN 2007; Sale-hyan 2007; Verhoeven 2011; IPCC 2012(a), 2013(a); Selby and Hoffmann 2014). For example, Salehyan argued that Ki Moon's attribution of climate as a root cause is irresponsible because it shifts liability and allows the government to blame the West. He considered that "Ban Ki-moon's case about Darfur was music to Khartoum's ears" and that it is preposterous to suggest that poor rainfall, instead of government and combatant action, caused the genocide. He concluded that "by Moons perverse logic, consumers in Chicago and Paris are at least as culpable for Darfur as the regime in Khartoum" (Salehyan 2007).

Similarly, Verhoeven described:

The regime loves the "climate war" rhetoric because it obscures its own role and, above all, Sudan's fundamental problem since independence: Khartoum's logic of rule is inextricably linked to exclusion, patronage and violence.

(Verhoeven 2011, 695)

Verhoeven expressed concern that the government has systematically used the climate-Darfur rhetoric to prevent discussions about inequalities, concluding that

Sachs and Ban Ki-Moon have sadly bought into the a-political argument about development remedying scarcity, failing to understand that Khartoum-led "development" is exactly what caused scarcity and violence in the first place.

(Ibid.)

These perspectives were also captured in the first and second drafts of the IPCC's assessment report, which stated that such attribution masks the culpability of actors and major drivers of insecurity (IPCC 2012(a), 2013(a)). However, they were removed from the final version of the report.

Figure 5.6 gives an overview of the actors, translations and knowledge resources relevant to this controversy.

Ambassador Mohamad cited Ki Moon's 2007 article as a source for the notion that Darfur was a climate conflict, and, on that basis, argued that there was nothing the government could do about it. Faris quoted Michael Byers, political scientist at the University of British Columbia, to highlight that the climate-Darfur link "changes us from the position of Good Samaritans – disinterested, uninvolved people who may feel a moral obligation – to a position where we, unconsciously and without malice, created the conditions that led to this crisis" (2007).

IRIN's criticism of Ki Moon cited Julie Flint, a journalist, who acknowledges that there is truth to climate-conflict linkages, but emphasized that simplification "whitewashes the Sudan government", and that the conflict was caused by government response to the rebellion (IRIN 2007). Referring to *How to Present the Next Darfur* in *Time* magazine on 26 April 2007, Verhoeven observed how pro-government lobbyists seized climate-conflict narratives, and quoted David Hoile's

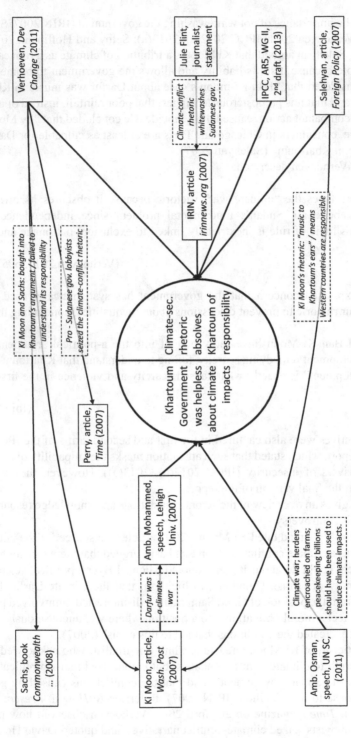

Figure 5.6 Actors, translations and knowledge resources related to implications of climate-Darfur links on Khartoum's responsibility for the conflict

2008 book *Darfur: The Road to Peace* to argue that "climate change and directly related environmental degradation have served to underpin the politics of conflict in Darfur" (Verhoeven 2011, 695).

This chapter has described the controversies, translations and knowledge resources involved in the climate-Darfur debate. The next chapter will consider the same questions for Syria's civil war that began in 2011.

Notes

1 www.bbc.com/news/world-africa-22336600
2 At: https://space.oscar.wmo.int/
3 The model simulations compared included:

- GFOL CM2.0 and GFOL CM2.1, by US Department of Commerce, NOAA, and Geophysical Fluid Dynamics Laboratory at Princeton.
- MIUB ECHO-G by Meteorological Institute of University of Bonn and Korea Meteorological Administration.
- MPI ECHAM5 by the Max Planck Institute for Meteorology.
- MRI CGCM23 by the Meteorological Research Institute, Japan.

Literature

Adelphi 2015: *Drought, Migration and Civil War in Darfur (ECC Factbook Conflict Analysis)*. Video published on YouTube on 28 October 2015. At: www.youtube.com/watch?v=_MF2ZAHDdoQ (accessed 5 March 2021).

Akasha, M.O. 2014: *Darfur – a Tragedy of Climate Change*. Anchor Academic Publishing, Hamburg.

Anderson, S. 2004: *How did Darfur happen?* The New York Times Magazine. 17 October. At: www.nytimes.com/2004/10/17/magazine/how-did-darfur-happen.html (accessed 24 January 2021).

Baldauf, S. 2006: *Africans are already facing climate change*. In: Christian Science Monitor. 6 November. At: www.csmonitor.com/2006/1106/p04s01-woaf.html (accessed 28 October 2019).

Bates, J.; Lin, Y. and Goodale, P. 2016: *Data journeys: Capturing the socio-material constitution of data objects and flows*. In: Big Data & Society, July–December, 1–12.

Biden, J. Jr. 2020: *Statement at the CNN Democratic Presidential Debate*. 15 March. At: www.youtube.com/watch?v=2z613_M5gxE (accessed 9 November 2020).

Braun, J. 2006: *A hostile climate – did global warming cause a resource war in Darfur?* SEED Magazine. 2 August. At: http://web.archive.org/web/20061103072503/www.seedmagazine.com/news/2006/08/a_hostile_climate.php?page=1 (accessed 28 September 2020).

Bromwich, B.; Adam, A.A.; Fadul, A.A.; Chege, F.; Sweet, J.; Tanner, V. and Wright, G.: *Darfur: Relief in a vulnerable environment*. Tearfund. March. At: https://reliefweb.int/sites/reliefweb.int/files/resources/7801DDA50C25BC2FC1257309002EA80F-Full_Report.pdf (accessed 3 November 2020).

Brown, I.A. 2010: *Assessing eco-scarcity as a cause of the outbreak of conflict in Darfur: A remote sensing approach*. In: International Journal of Remote Sensing, 31(10), 2513–20.

Burke, M.B.; Miguel, E.; Satyanath, S.; Dykema, J.A. and Lobell, D.B. 2009: *Warming increases the risk of civil war in Africa*. In: PNAS, 106(49), 20670–4.

Cameron, D. 2007: *A warmer world is ripe for conflict and danger*. Financial Times, 24 January, 15.IA Corporation 2007: *National Security and the Threat of Climate Change*. CNA Corporation, Washington, DC.

Degomme, O. and Guha-Sapir, D. 2010: *Patterns of mortality rates in Darfur conflict*. In: Lancet, 375, 249–300.

De Waal, A. 2007(a): *Is climate change the culprit for Darfur?* African Arguments. 25 June. At: https://africanarguments.org/2007/06/25/is-climate-change-the-culprit-for-darfur/ (accessed 6 November 2020).

De Waal, A. 2007(b): *Reply to cause and effect*. African Arguments. 8 August. At: https://africanarguments.org/2007/08/cause-and-effect/ (accessed 13 February 2021).

Edwards, P.N. 2013: *A Vast Machine – Computer Models, Climate Data, and the Politics of Global Warming*. MIT Press, Cambridge, MA.

Faris, S. 2007: *The real roots of Darfur*. Atlantic Monthly. April 2007. At: www.theatlantic.com/magazine/archive/2007/04/the-real-roots-of-darfur/305701/ (accessed 21 October 2019).

Flint, J. and De Waal, A. 2008: *Darfur: A New History of a Long War*. African Arguments. Zed Books, London.

Giannini, A.; Saravanan, R. and Chang, P. 2003: *Oceanic forcing of Sahel rainfall on interannual to interdecadal time scales*. In: Science, 302, 1027–30.

GWR 2019: *First Climate Change War*. At: www.guinnessworldrecords.com/world-records/first-climate-change-war (accessed 7 September 2021).

Hagan, J. and Kaiser, J., 2011: *The displaced and dispossessed of Darfur: Explaining the sources of a continuing state-led genocide*. In: British Journal of Sociology, 62(1), 1–25.

Homer-Dixon, T. 2007: *Cause and Effect*. In: SRC Blogs – Climate and Environment, Making Sense of Sudan. 2 August. At: https://homerdixon.com/cause-effect/ (accessed 5 June 2020).

Hughes, H. 2007: *Climate Change and Securitization*. Unpublished Master's thesis, Cambridge University, Cambridge. At: www.academia.edu/4591407/Climate_Change_and_Securitization (accessed 24 July 2022).

Hulme, M. 1999: *'gu23wld0098.dat', Version 1.0, March 1999*. UEA CRU. At: https://crudata.uea.ac.uk/cru/data/precip/gu23wld0098_doc.pdf (accessed 10 November 2020).

Ingham, R. and Hood, M. 2008: *Climate change: Progress at polluters' talks, but obstacles ahead*. Sydney Morning Herald. 19 April. At: www.smh.com.au/world/climate-change-progress-at-polluters-talks-but-obstacles-ahead-20080418-273v.html (accessed 9 November 2020).

IPCC 2007 (Trenberth, K.E.; Jones, P.D.; Ambenje, P.; Bojariu, R.; Easterling, D.; Klein, Tank A.; Parker, D.; Rahimzadeh, F.; Renwick, J.A.; Rusticucci, M.; Soden B. and Zhai, P. 2007: *Observations: Surface and atmospheric climate change*. In: Solomon, S.; Qin, D.; Manning, M.; Chen, Z.; Marquis, M.; Averyt, M.B.; Tignor, M. and Miller, H.L. (Eds.): Climate Change 2007: The Physical Science Basis – Contribution of Working Group I to the Fourth Assessment Report of the IPCC. Cambridge University Press, Cambridge and New York, NY.

IPCC 2012(a): *Working Group II, 5th Assessment Report, Chapter 12: Human Security. First-Order Draft*. 11 June. At: https://archive.ipcc.ch/pdf/assessment-report/ar5/wg2/drafts/WGIIAR5-Chap12_FOD.pdf (accessed 5 June 2020).

IPCC 2012(b): *Expert Review of Working Group II, 5th Assessment Report, Chapter 12: Human Security. First-Order Draft*. 6 August. At: https://archive.ipcc.ch/pdf/assessment-report/ar5/wg2/drafts/WGIIAR5_FODCh12_annotation.pdf (accessed 5 June 2020).

IPCC 2013(a): *Working Group II, 5th Assessment Report, Chapter 12: Human Security. Second-Order Draft.* 28 March. https://archive.ipcc.ch/pdf/assessment-report/ar5/wg2/drafts/WGIIAR5-Chap12_SOD.pdf (accessed 5 June 2020).

IPCC 2013(b): *Expert Review of Working Group II, 5th Assessment Report, Chapter 12: Human Security. Second-Order Draft.* 24 May. At: https://archive.ipcc.ch/pdf/assessment-report/ar5/wg2/drafts/WGIIAR5_SODCh12_annotation.pdf (accessed 5 June 2020).

IPCC 2013(c): *Working Group II, 5th Assessment Report, Chapter 12: Human Security. Final Draft.* 28 October. https://archive.ipcc.ch/pdf/assessment-report/ar5/wg2/drafts/fd/WGIIAR5-Chap12_FGDall.pdf (accessed 5 June 2020).

IPCC 2013(d): *Procedures for the Preparation, Review, Acceptance, Adoption, Approval and Publication of IPCC Reports.* At: https://archive.ipcc.ch/pdf/ipcc-principles/ipcc-principles-appendix-a-final.pdf (accessed 5 June 2020).

IPCC 2014 (Adger, W.N.; Pulhin, J.M.; Barnett, J.; Dabelko, G.D.; Hovelsrud, G.K.; Levy, M.; Oswald Spring, Ú. and Vogel, C.H.): *Human security.* In: Field, C.B.; Barros, V.R.; Dokken, D.J.; Mach, K.J.; Mastrandrea, M.D.; Bilir, T.E.; Chatterjee, M.; Ebi, K.L.; Estrada, Y.O.; Genova, R.C.; Girma, B.; Kissel, E.S.; Levy, A.N.; MacCracken, S.; Mastrandrea, P.R. and White, L.L. (Eds.): Climate Change 2014: Impacts, Adaptation, and Vulnerability. Part A: Global and Sectoral Aspects. Contribution of Working Group II to the Fifth Assessment Report of the IPCC. Cambridge University Press, Cambridge and New York, NY, 755–91.

IRIN 2007: *Sudan: Climate change – Only one cause among many for darfur conflict.* 28 June. At: www.globalsecurity.org/military/library/news/2007/06/mil-070628-irin03.htm (accessed 9 November 2020).

Kevane, M. 2009: *Two graphs of rainfall and temperature in Darfur.* In: Understanding Sudan: Commentary. 3 November. At: http://sudancommentary.blogspot.com/2009/11/two-graphs-of-rainfall-and-temperature.html (accessed 26 July 2022).

Kevane, M. and Gray, L. 2008: *Darfur: Rainfall and conflict.* In: Environmental Research Letters, 3. At: https://iopscience.iop.org/article/10.1088/1748-9326/3/3/034006/pdf (accessed 21 November 2019).

Ki Moon, B. 2007: *A climate culprit in Darfur.* Washington Post. 16 June. At: www.washingtonpost.com/wp-dyn/content/article/2007/06/15/AR2007061501857.html (accessed 21 October 2019).

Klare, M.T. 2006: *The coming resource wars.* AlterNet. 10 March. At: www.alternet.org/2006/03/the_coming_resource_wars/ (accessed 21 November 2019).

Law, J. and Mol, A. 2001: *Situating technoscience: An inquiry into spatialities.* In: Environment and Planning D: Society and Space, 19(5), 609–21.

Mamdani, M. 2009: *Saviors and Survivors: Darfur, Politics, and the War on Terror.* Doubleday, New York.

Mazo, J. 2010: *Climate Conflict.* Routledge, New York.

Miller, C. 2004: *Climate science and the making of a global political order.* In: Jasanoff, S. (Ed.) 2004, 46–66.

Mjøs, O.D. 2007: *Nobel Peace Prize 2007 Award Ceremony Speech.* Copyright The Nobel Foundation. 10 December. At: www.nobelprize.org/prizes/peace/2007/ceremony-speech/ (accessed 9 November 2020).

Moran, S. 2020: *Fact check: Joe Biden says climate change caused Darfur conflict.* Breitbart News. 15 March. At: www.breitbart.com/politics/2020/03/15/fact-check-joe-biden-says-climate-change-caused-darfur-conflict/ (accessed 6 November 2020).

Reynolds, P. 2007: *Security council takes on global warming.* BBC News. 18 April. At: http://news.bbc.co.uk/2/hi/americas/6559211.stm (accessed 5 June 2020).

Russell, B. and Morris, N. 2006: *Armed forces are put on standby to tackle threat of wars over water*. The Independent. 28 February. At: www.independent.co.uk/environment/armed-forces-are-put-on-standby-to-tackle-threat-of-wars-over-water-6108139.html (accessed 28 September 2020).

Sachs, J. 2006: *Ecology and political upheaval*. In: Scientific American. 1 July. At: www.scientificamerican.com/article/ecology-and-political-uph/ (accessed 21 November 2019).

Salehyan, I. 2007: *The new myth about climate change*. In: Foreign Policy. 14 August. At: https://foreignpolicy.com/2007/08/14/the-new-myth-about-climate-change/ (accessed 3 November 2020).

Schwartz, P. and Randall, D. 2003: *An Abrupt Climate Change Scenario and Its Implications for United States National Security*. Washington, DC. At: https://training.fema.gov/hiedu/docs/crr/catastrophe%20readiness%20and%20response%20-%20appendix%202%20-%20abrupt%20climate%20change.pdf (accessed 6 August 2021).

Selby, J. and Hoffmann, C. 2014: *Beyond scarcity: Rethinking water, climate change and conflict in the Sudans*. In: Global Environmental Change, 29, 360–70.

Srinivasan, S. and Watson, E.E. 2013: *Climate change and human security in Africa*. In: Redclift, M.R. and Marco, G. (Eds.): *Handbook on Climate Change and Human Security*. Edward Elgar Publishing Ltd., Cheltenham, UK.

Stern, N. 2006: *The Economics of Climate Change: The Stern Review*. Cambridge University Press. Cabinet Office – HM Treasury. Crown copyright, Cambridge.

Straw, B. 2007: *Sudanese ambassador: Darfur is a classic case of climate change*. Lehigh University News. 16 November. At: https://www2.lehigh.edu/news/sudanese-ambassador-darfur-classic-case-climate-change (accessed 9 November 2020).

Sunga, L.S. 2011: *Does climate change kill people in Darfur?* In: Journal of Human Rights and the Environment, 2(1), 64–85.

UNEP 2007: *Sudan – Post-conflict Environmental Assessment*. At: https://postconflict.unep.ch/publications/UNEP_Sudan.pdf (accessed 21 November 2019).

UNSC 2007(a): *Security Council Holds First-Ever Debate on Impact of Climate Change on Peace, Security, Hearing Over 50 Speakers*. UNSC Press Release, SC/9000. 17 April 2007. At: www.un.org/press/en/2007/sc9000.doc.htm (accessed 20 August 2020).

UNSC 2007(b): *Letter Dated 5 April 2007 from the Permanent Representative of the United Kingdom of Great Britain and Northern Ireland to the UN Addressed to the President of the Security Council*. S/2007/186. 5 April. At: www.securitycouncilreport.org/atf/cf/%7B65BFCF9B-6D27-4E9C-8CD3-CF6E4FF96FF9%7D/Ener%20S%202007%20186.pdf (accessed 29 September 2020).

UNSC 2011: *Record of the 6587th Meeting*. Document S/PV.6587. At: https://undocs.org/en/S/PV.6587 and https://undocs.org/en/S/PV.6587(Resumption1) (accessed 20 December 2020).

University for Peace Africa Programme (UPAP) and University of Khartoum, Peace Research Institute (PRI) 2006: *Environmental Degradation as a Cause of Conflict in Darfur: Conference Proceedings. Khartoum*. December. At: www.worldcat.org/title/environmental-degradation-as-a-cause-of-conflict-in-darfur-conference-proceedings-khartoum-december-2004/oclc/124517641 (accessed 13 February 2021).

Verhoeven, H. 2011: *Climate change, conflict and development in Sudan: Global neo-Malthusian narratives and local power struggles*. In: Development and Change, 42(3), 679–707.

Welzer, H. 2012: *Climate Wars: What People Will Be Killed for in the 21st Century*. Polity Press, Cambridge.

WMO 2014(a): *Volume A – Observing Stations*. Copyright WMO, Geneva. At: https://extranet. wmo.int/pages/prog/www/ois/Operational_Information/VolumeA/VolumeA2014ed.pdf (accessed 16 October 2020).

WMO 2014(b): *Volume A – Observing Stations, Code Tables for the Master Flat File*. At: www.wmo.int/pages/prog/www/ois/volume-a/9ACodeTables9805.html (accessed 10 Nov 2020).

Additional literature referenced by the contributions to the climate-Darfur debate

De Waal, A. 1989: *Famine that Kills: Darfur, Sudan, 1984–1985*. Clarendon, Oxford.

Held, I.M.; Delworth, T.L.; Lu, J.; Findell, K.L. and Knutson, T.R. 2005: *Simulation of Sahel drought in the 20th and 21st century*. In: PNAS, 102(50), 17891–6.

Hoile, D. 2008: *Darfur: The Road to Peace*. European-Sudanese Public Affairs Council, London.

ICG 2002: *God, Oil and Country – Changing the Logic of War in Sudan*. Africa Report No. 839. International Crisis Group Press, Brussels.

Mahe, G., L'Hote, Y; Olivry, J.C. and Wotling, G. 2001: *Trends and discontinuities in regional rainfall of West and Central Africa: 1951–1989* In: Hydrological Science Journal, 46, 211–26.

Miguel, E.; Stayanath, S. and Sergenti, E. 2004: *Economic shocks and civil conflict: An instrumental variables approach*. In: Journal of Political Economy, 112, 725–53.

Nicholson, S.E. 2001: *Climatic and environmental change in Africa during the last two centuries*. In: Climate Research, 17, 123–44.

Patey, L. 2010: *Crude days ahead? Oil and the resource curse in Sudan*. In: African Affairs, 109, 617–36.

Pettitt, A.N. 1979: *A non-parametric approach to the change-point problem*. In: Applied Statistics, 28, 126–35.

Prince, S.D.; Wessels, K.J.; Tucker, C.J. and Nicholson, S.E. 2007: *Desertification in the Sahel: a reinterpretation*. In: Global Change Biology, 13, 1308–13.

Suliman, O. 2010: *The Darfur Conflict: Geography or Institutions?* Routledge, London.

Teklu, T. and Von Braun, J. 1991: *Drought and Famine Relationships in Sudan: Policy Implications*. IFPRI, Washington DC.

Thornton, P.K.; Jones, P.G.; Owiyo, T.; Kruska, R.L.; Herrero, M.; Kristjanson, P.; Notenbaert, A.; Bekele, N. and Omolo, A., with contributions from Orindi, V.; Otiende B.; Ochieng, A.; Bhadwal, S.; Anantram, K.; Nair, S.; Kumar, V. and Kulkar, U. 2007: *Mapping Climate Vulnerability and Poverty in Africa*. Report to DFID, ILRI, Nairobi, Kenya. At: https://media.africaportal.org/documents/Mapping-Vuln-Africa_bpuNkPK.pdf (accessed 5 June 2020).

Young, H. and Osman, A.M.K. 2005: *Darfur: Livelihoods under Siege*. Tufts University, Feinstein International Famine Center, Medford, MA.

6 Syria

Did climate change "open the gates of hell"?

The previous chapter described the translations of climate-Darfur links conflict and the role of knowledge resources in those translations. This chapter describes, in similar terms, the making and unmaking of climate-conflict links in the context of the Syrian civil war that began in 2011. The first section provides a brief history of the conflict, followed by a chronological synthesis of the climate-Syria debate. The next section will identify five key controversies related to climate-Syria links. Then, based on those controversies, the sections that follow will describe, for each controversy, how actors translate climate and conflict together and mobilize knowledge resources.

Conflict background

There is agreement that the Syria conflict followed the "wave of optimism" after the Arab Spring and toppling of dictators in Tunisia, Egypt and Libya. These and local events led to similar demands in Syria (Phillips 2020, viii). On 12 March, Kurds protested in the East, and on 15 March, groups of Syrians expressed solidarity with the Arab Spring at Egyptian and Tunisian embassies in Damascus and residents of the southern town of Dara'a demanded the release of local youths who had been arrested and tortured for anti-Assad graffiti (Fröhlich 2016). In response, security forces killed four protesters, triggering larger protests, now also directed at the regime and the secret services (Phillips 2020, 45). The regime responded with more violence, but news of the events spread, and protests began in other cities, including Latakia, Tarsus, Idlib, Qamishli, Deir-ez Zor, Raqqa and Hama (Ibid., 49–50).

The regime labeled the protesters "armed gangs" or jihadists, employed agent provocateurs to shoot at soldiers to justify repression, and organized counter-protests (Ibid., 53–5). Assad promised investigations of events in Dara'a, release of political prisoners, a new cabinet, reduced emergency laws, elections and a national dialogue (Ibid., 55, 66 and 84). But the violence had already motivated a broader insurgency that demanded Assad's fall. Local militias started to protect protests and later to fight the regime, and army deserters joined the opposition. By the summer of 2011, Syria was in a state of civil war (Ibid., 2).

DOI: 10.4324/9781003451525-6

Initially, western countries urged restraint and negotiations (Ibid., 65), and increased sanctions (Ibid. 77). However, the violent "Ramadan assaults" against protesters in Hama, Homs and Deir-ez-Zor in July–August 2011 prompted calls for Assad's removal, including in an announcement by the United States, the UK, Germany, France and Canada on 18 August (Ibid., 79). Assad's allies Qatar and Turkey abandoned him, and began, along with Saudi Arabia, to arm rebel groups (Ibid., 125). However, Russia and Iran declared their support to Assad (Ibid., 60). This constellation motivated further violence, as both sides of the conflict felt reassured by the clear signals of support from foreign partners (Ibid. 82).

While the opposition was initially decentralized (Ibid., 106), in late July 2011, major groups were organized into a coalition named the Free Syrian Army (FSA) (Ibid., 84–5), which, however, faced coordination and legitimacy challenges amid diverging interests, ideologies and capacities of its members (Ibid., 127). On 23 August, the Syrian National Council was formed in Istanbul (Ibid., 72). It brought together the Muslim Brotherhood and local committees, with the explicit goal of overthrowing the regime (Ibid., 106). However, it quickly lost legitimacy, and in November 2012 was folded into the National Coalition for Syrian Revolutionary and Opposition Forces (Ibid., 115). Later, a Supreme Military Command was established to coordinate the other rebel groups (Ibid., 129).

Toward the end of 2011, the Arab League proposed a peace plan with international monitoring, which Assad agreed to. However, the monitors were few, unprepared for the operation and lacked support of key governments. The mission broke down in January 2012 (Ibid., 88–91). Afterward, the Gulf countries tried to bring Syria up in the UNSC, but were thwarted by Russia and China, who would veto any resolutions calling for regime change, condemnation of violence, sanctions or interventions (Ibid., 92).

In March, the "Annan plan", unanimously supported by the UNSC, called for a cease-fire, UN supervision, humanitarian help, release of prisoners, and dialogue. A ceasefire began on 12 April and a 300-person UN Supervision Mission in Syria was dispatched. However, massacres continued, and the monitors were attacked. The mission ended in August. After this, Annan convened the Action Group for Syria to reduce hostilities by bringing together international backers of the warring parties. However, key actors, such as Iran, were excluded, and the priorities of participants diverged: the United States saw the group as a "blueprint for Assad's departure" – a red flag for Russia. Neither the regime nor the opposition took the plan seriously, and Annan resigned (Ibid., 99–102).

Initially, Assad's military had superior weaponry and air power (Ibid., 126). However, the rebels secured weapons from captured bases and international supporters (Ibid., 127). In 2012, they captured military bases outside urban centers and forced the regime to withdraw from the east and the north and to focus on western Syria and key gas fields. In July the rebels even attacked Damascus (Ibid., 128). Toward the end of 2012, the regime appeared "on the verge of collapse" (Ibid., 148). However, a badly coordinated rebel attack on Aleppo in July led to a stalemate, a division of the city into two parts, and loss of rebel momentum (Ibid., 129).

Parallel to these developments, the presence of Islamist groups grew, motivated partly by FSA failures, the "warlordism" and lack of discipline of some rebels, and foreign support (Ibid., 131). Some groups (including Tawheed, Farouq, Syrian Islamist Liberation Front, Syrian Islamic Front, Ahrar al-Sham and Liwa al-Haqq) waged a domestic jihad against Assad, but two others, Jabhat al-Nusra and ISIS, considered Syria as part of a global jihad. In 2013, ISIS began to capture locations in the east, including Raqqa, Deir ez-Zor and the Tabqa dam, and proclaimed itself as the Islamic State in 2014 (Daoudy 2019, 95; Phillips 2020, 196). After this, ISIS was at war with nearly everyone else, though it was less focused on overthrowing Assad than consolidating a caliphate in northern Iraq and eastern Syria.

Around the same time, Kurdish militias organized under the YPG (People's Protection Units), a PKK-affiliated group, and the Kurdish National Council, a local-oriented pacifist group. As ISIS had done, the YPG filled parts of the power vacuum resulting from the withdrawal of Assad's forces from north and east in 2012, leading to clashes with the FSA and major battles with ISIS (Ibid., 134). In November 2013, the YPG proclaimed Western Kurdistan (Rojava) (Daoudy 2019, 95; Phillips 2020, 186), setting the stage for later confrontation with Turkey.

Each conflict party was backed by foreign governments or organizations, who pursued their own goals (Ibid., 3), with conflict resolution as a secondary objective (Ibid., 102). And while each party received enough support to continue fighting, none received enough for a decisive victory (Ibid., 7 and 126). Assad was supported by Russia and Iran. Kurdish and moderate opposition were backed by the United States, the UK and France (Phillips 2020, 7). Saudi Arabia, Qatar and Turkey supported Islamist groups (Daoudy 2019, 9). Lebanon, Israel, the UAE and China also played a limited role (Phillips 2020, 7).

Starting in 2013, the scales began to tip in Assad's favor. In August 2013, a chemical attack in Ghouta killed 1,400 people. The United States had identified chemical weapons as a "red line". However, instead of a US intervention, Russia mediated a deal under which Assad would destroy his chemical arsenal and allow UN supervision. This dashed rebel hopes of a US operation against Assad (Ibid., 168–9, 177–81). A parallel turning point was the involvement of Iran and its allies in 2013–2014. Coordinated support by IRCG, Hezbollah and Shia militias enabled the regime to consolidate control of western Syria and block the rebels in Aleppo (Ibid., 151). Building on the chemical weapons deal, peace talks continued in January 2014 in Geneva. However, various rebel groups refused to meet each other, and the regime, having the military upper hand, showed no interest (Ibid., 190–1).

A decisive point was the Russian intervention in the summer of 2015 (Ibid., 9), likely triggered by the threat of ISIS, concerns about US hegemony, and Assad's military setbacks. Russia deployed its Black Sea fleet, 28 aircraft, and 2,000 personnel to support the regime's efforts to regain territory (Ibid., 217). Russia stated it was targeting ISIS, but it mostly attacked the anti-Assad forces (Ibid., 218). By early 2016, the military balance was in the regime's favor (Ibid., 219), and in December 2016, Assad retook eastern Aleppo (Ibid., 240). This forced rebel groups into negotiations, and new talks began in Vienna in November 2016, resulting in a cease-fire and establishment of four de-escalation zones announced by Iran, Russia and Turkey in February 2017.

After the Russian intervention, the Syrian opposition never recovered. Throughout 2017, various groups focused on ISIS. Kurdish forces, backed by US air power, rolled back the ISIS main forces from the north, and captured Raqqa in October (Ibid., 211, 255). In parallel, Assad's forces, supported by Russian bombardment and special forces, as well as by Iran-affiliated militias, captured ISIS territory from the west. The defeat of ISIS and the fact that Kurdish forces were less focused on fighting the regime than on defending Rojava from Turkish attacks enabled Assad to deal with remaining pockets of opposition. Beginning in 2018, regime forces began to pick the four de-escalation zones one by one (Ibid., 263). This followed a pattern: the regime would surround and shell a de-escalation zone, forcing rebels there to surrender. Fighters refusing reconciliation would be shipped to the Idlib region. Soon all de-escalation zones except Idlib were under regime control.

As of 2023, rebels still held the Idlib province under regime bombardment. Kurdish forces operate in the northeast, but Turkey has established buffer zones against them. The war has killed approximately 500,000 Syrians, forced five million to flee the country (mainly to Turkey, Jordan and Lebanon), and displaced 6.6 million internally. This has been accompanied by a total collapse of national health care, economy, culture and life expectancy (Ibid., 1).

The conflict has demonstrated the limits of US power to change things, and the capacity of other powers (in particular Russia and Iran) to maintain the status quo. The refugee crisis has fueled populist campaigns across the west, and UN conflict resolution have been incapable of making a difference (Ibid., viii). While still not over, Syria clearly is a war "that everyone lost" (Ibid., 304).

The climate-Syria debate

The first publication to connect climate and Syria was *Syria: Climate Change, Drought and Social Unrest*, by Werrell and Femia, published in February 2012 on *climateandsecurity.org*, website of the CCS. They described how Syria's 2006–2011 drought ruined agriculture and livelihoods of 800,000 Syrians, left one million with food insecurity, and triggered an exodus to cities. The authors recognized the many factors of the war, but highlighted how NOAA linked the drought to climate change, and how climate models project the loss of 29–57 per cent of rain-fed crops in 2010–50.

Shortly afterward, in April 2012, *New York Times* (*NYT*) journalist Thomas Friedman argued in *The Other Arab Spring* (2012) that the "Arab awakening" was driven by political and economic, but also climate stresses, and that stabilization efforts must focus on the latter. He also stipulated that in 2006–2011, total 60 per cent of Syria's land saw worst droughts and crop failures in history.

In August 2012, the *Bulletin of the Atomic Scientists* published Mohtadi's *Climate Change and the Syrian Uprising*. He argued that the role of climate has been unnoticed, described the drought's severity, and noted that while one drought cannot be attributed to climate, NOAA stated that natural variability cannot explain this one, and that GHGs accounted for half of the additional dryness in 1902–2010. He claimed that the drought displaced one and a half million people to urban peripheries, and caused an agricultural collapse, which Syria had no system to deal with.

Also in August 2012, Robin Mills, head of consulting at Manaar Energy, in *The National*, a newspaper based in the United Arab Emirates, highlighted drought severity and how it correlated with predictions of global warming – induced drying. Mills outlined that by 2011, one million Syrians were hungry and two to three million in poverty, and global food prices worsened things. He concluded that climate change can drive instability in water-scarce regions and overburden weak regimes.

In May 2013, Thomas Friedman published *Without Water, Revolution* in the *NYT*. He described how a combination of the worst drought in Syria's history, population growth, repression and outside influence caused the conflict, and that climate change will lead to more. He discussed how the drought ruined the livelihoods of 800,000 Syrians and triggered a mass migration to cities, and how Assad's neglect of the refugees led to unemployment and radicalization – "a prescription for revolution".

In September 2013, the climate-Syria link was discussed in National Public Radio (NPR) broadcast *How could drought spark a civil war?* Host Jacki Lyden and guest Nayan Chanda described the drought of 2006–2010, its consequences, the rural-urban migration of one and a half million Syrians, and how these factors "created a situation which started the first spark" in the city of Dara'a (NPR 2013).

A 2014 report by the CNA Military Advisory Board, *National Security and the Accelerating Risks of Climate Change*, stipulated that the conflict was preceded by drought, which, together with inadequate response and overgrazing, decimated livestock and crops and forced millions to migrate to cities. Antigovernment forces were emboldened. The report suggested that a better understanding of "cascading climate-related impacts" can help avoid future conflicts (CNA 2014, 13–14).

The first research article considering climate-Syria links was published by Francesca De Châtel in *Middle Eastern Studies*. She considered that while it is difficult to prioritize conflict triggers, the uprisings were caused by several factors, including poverty, liberalization, subsidy cuts, corruption, unemployment and drought. She argued that overstating the role of climate distracts from core problems and enables the Assad regime to blame external factors for the conflict (2014).

In April 2014, *Showtime* launched the Emmy-winning documentary series *Years of Living Dangerously*. It was produced by James Cameron and Arnold Schwarzenegger, involved James Hansen and Michael Mann as science advisors, and starred Harrison Ford, Sigourney Weaver and President Obama and others. In the first episode, *Dry Season*, Friedman investigated the climate-Syria link (2014). He reviewed studies; interviewed refugees, fighters and US National Security Adviser Rice; and traveled to Turkey and Syria. He concluded that the volatile region is getting hotter, and we must take notice.

In June 2014, *Slate* published Eric Holthaus' *Is Climate Change Destabilizing Iraq?* Holthaus described how global warming made droughts a fixed condition in the Middle East, and how the drought of 2006–2010 assisted destabilization, along with other drivers, by forcing farmers to abandon fields and to "flood cities with protests". He considered that climate might have contributed to the rise of ISIS, given that 2014 was very hot, and that studies link higher temperatures and violence.

Similarly, on 30 June, in a France 24 report *ISIS and Climate Change* (2014), Florence Villeminot cited Holthaus' article to highlight that climate change might have contributed to ISIS' success, as the militants jumped into the chaos caused by the unusual 2014 heat and drying of agricultural lands, and that research has linked heat and instability. The climate-ISIS link was also explored by Kelly Berkell (2014) in September in the *Huffington Post* piece *How Climate Change Helped ISIS*. Berkell argued that ISIS "stole onto the scene" in the chaos triggered by the climate-induced drought, which caused a humanitarian crisis and fueled the uprisings.

In July 2014, the journal *Weather, Climate and Society* published Peter Gleick's *Water, Drought, Climate Change, and Conflict in Syria* (2014). Gleick assessed the connections between water and the conflict. He described the war as a result of multiple factors, but emphasized that "water and climatic conditions have played a direct role in the deterioration of Syria's economic conditions" (331). He argued that since conflicts do not have single causes, conflict studies should consider multiple factors and links between them.

In the fall of 2014, Femia and Werrell, together with Troy Sternberg (Femia et al. 2014), published their second contribution in the *Seton Hall Journal of Diplomacy and International Relations*. They investigated how the long-term drought, displacement and lack of government response may have contributed to the conflict. They considered the drought severity, its economic and migration consequences, its links with climate change, and the role of rural migrants in urban uprisings.

In early 2015, authors led by Colin Kelley at the University of California, Santa Barbara, published *Climate change in the Fertile Crescent and implications of the recent Syrian drought* in PNAS (Kelley et al. 2015). Based on rainfall data, satellite observations, vegetation and water indices, tree ring studies, information from humanitarian organizations, climate models, and statistical and presentational methods, the authors considered that Syria experienced its worst drought on record in 2006–10, which contributed to unrest. They observed drying and warming trends, for which they found no natural causes, but which correspond to models that reflect increased GHG emissions. They concluded that the pre-war drought has become more than twice as likely due to climate change.

In March 2015, Richard Gray (2015) published *Did Climate Change Trigger the War in Syria?* in The *Daily Mail*. He described Kelley et al. (2015) as "the first study looking at the role of climate change in the Syrian war", highlighting their findings about the worst drought ever, likely strengthened by climate change, loss of agricultural lands, rural-urban migration, and subsequent poverty and unrest that triggered the war in 2011. Similarly, on 2 March 2015, *Scientific American* published Mark Fischetti's (2015) *Climate Change Hastened Syria's Civil War*. Fischetti also reported on Kelley et al., and reiterated that a drought, made more likely by climate change, combined with other stressors and pushed things "over the threshold". He interviewed Richard Saeger, one of the authors of Kelley et al., who stated that the entire Middle East faces similar risks. Mark Zastrow's *Climate Change Implicated in Current Syrian Conflict* (2015), published in March by *Nature*, also highlighted Kelley et al., in particular that the drought was part of a long-term trend, consistent with anthropogenic change. He interviewed Francesco Femia (CCS), who described the dynamics of migration and unrest, while highlighting difficulties with attributing violence to drought, and the risks of focusing on climate as a conflict factor. The climate-Syria-ISIS link was also discussed by Russell Brand (2015) in *ISIS versus Climate Change – Which Kills More?*, published on YouTube in March 2015. Brand recapped Kelley et al.'s findings, and added that climate-induced hunger helped ISIS recruitment, and that global corporations and capitalism, being responsible for climate change, are also responsible for the Syrian conflict.

In May 2015, in a speech at the US Coast Guard Academy, President Obama highlighted, when discussing climate change, that "it's now believed that drought and crop failures and high food prices helped fuel the early unrest in Syria, which descended into civil war" (Obama 2015).

In June 2015, the SAIS Review published *Did We See It Coming? State Fragility, Climate Vulnerability, and the Uprisings in Syria and Egypt* by Werrell, Femia and Sternberg (Werrell et al. 2015). They described how drought and resource management contributed to Syria's instability, emphasizing drought severity, migration and the role of climate. They found that climate factors had been inadequately integrated into the Failed States Index and the Notre-Dame Global Adaptation Index, and that improvements are needed in factors, sources, weighing, and connections between indicators, in particular how climate vulnerability relates to state fragility.

Ian Bremmer, head of political risk consultancy Eurasia Group, argued in a Channel 4 interview in June 2015 that "the Syria war began because of climate change", as farmers had to kill livestock due to drought and lack of water, and then Assad "whacked them hard", "the country exploded" and turned into a failed state where ISIS "could come in" (Bremmer 2015).

At the 2015 Summit of the Americas, former US Vice President Al Gore described climate as 'the underlying story of what caused the gates of hell to open in Syria. He identified water shortage as a key factor, and cited familiar figures: in 2006–2010, Syria had a climate-related drought that ruined 60 per cent of farms, killed 80 per cent of livestock, and drove one and a half million people into cities, where they collided with one and a half million Iraqi refugees. He also quoted a US embassy cable published by WikiLeaks, which described Syrian ministers saying about the drought: "we can't deal with this" (Gore 2015).

In November 2015, YaleClimateConnections (YCC) published a five-minute video *Drought, Water, War, and Climate Change* on YouTube (2015). It cited, among others, Friedman's description of how the NOAA study on Mediterranean drought identified Syria "at the epicenter", suffering its worst drought before the revolution. On the video, CNN (Cable News Network)-footage described how climate may have fueled the war, with the 2006–10 drought displacing one and a half million people from farms to overcrowded cities.

In November 2015, Adelphi (2015) published on video *Climate Change and the Syrian Civil War*. It introduced a "conceptual model" identifying climate-Syria links, in particular a Mediterranean drying, which, according to NOAA, could not be explained by natural variability. The video described how the 2006–2010 drought, compounded by bad water management, drove two to three million people to poverty, causing an exodus to cities. This increased competition for housing, jobs and services; while food price shocks, corruption, Arab Spring revolutions, subsidy cuts, and lack of government help eroded the social contract.

Another video was published in December 2015 by *The Economist* (2015). *Warriors and Weather: Climate Change and National Security in America* included a section on Syria, in which US Navy Admiral David Titley (ret.) described how climate change can be a part of a chain of events leading to conflict, using Syria's drought as an example. As other conflict factors, Titley highlighted agricultural self-sufficiency policies, over-use of water and the presence of Iraqi refugees. These drove 750,000 farmers into cities, triggering an insurgency and creating a "fertile ground for ISIS".

A second critical contribution appeared in April 2016 in *Contemporary Levant*. Christiane Fröhlich, with the Institute for Peace Research and Security Policy of the University of Hamburg and the Istanbul Policy Center of the Sabanci University, sought to understand the role of drought-driven migration in the 2011 uprisings based on social movement theory and interviews with refugees. She concluded that environmental factors were only one driver of migration, that drought migrants had a limited role in the uprisings and were not integrated in the host communities to a degree that would enable systematic mobilization, and that the uprisings were mainly a result of regime oppression (2016).

As part of *The Years Project*, Friedman published a video *Climate Wars – Syria* in February 2017 (Friedman 2017), which replicated his segments from *Years of Living Dangerously* (Friedman 2014).

Selby et al. (2017(a)), in *Political Geography*, examined three assumptions about the climate-Syria link: (1) GHG emissions contributed to the drought; (2) drought triggered mass migration; and (3) migration pressure led to war. They criticized the limited use of rainfall data, varying estimates of drought duration, and use of linear rainfall trends, concluding that evidence does not prove drying of Syria. On migration, they considered refugee numbers in previous studies as too high or based on wrong data, and highlighted the omission of other causes of migration. In terms of migration-conflict link, the authors criticized the lack of references to conflict literature or ethnographic data and described how Fröhlich's interviews revealed no participation of refugees in uprisings or drought-related grievances in protests.

In the same issue of *Political Geography*, Kelley et al. 2017 argued that Selby et al. show no evidence that the pre-conflict conflict dryness was natural; that other evidence confirms regional climate impacts; that long-term drying agrees with human-induced climate change, climate models and a theory of winter circulation and climate; and that models might underestimate drying. The authors highlighted how population growth, poor policies and Iraqi refugees burdened resources, followed by the drought-induced agricultural collapse and migration. They argued that Selby et al. misinterpret the lack of evidence of drought-related grievances among protesters as proof of missing links, use migration numbers selectively, and ignore other proxy numbers as well as the migration that happened shortly prior to the uprisings. Kelley et al. concluded that more examination is needed because drying is projected to continue and has potential to cause more unrest, and that Selby et al. do not refute that.

Hendrix (2017), also in *Political Geography* 60, welcomed Selby et al.'s questioning of causal links and reminded of the importance of identifying pathways through which climate can trigger violence. He highlighted that climate-conflict links are context-dependent, worried about the overly vocal contributions of non-specialists and the public, and emphasized that findings about the link tend to be probabilistic. Hendrix recommended studying how climate interacts with contexts to result in violence, expanding the geographical reach of studies, and avoiding causal language when evidence supports probabilistic climate-conflict links, rather than using evidence to explain a particular event, and making it sound like a necessary condition that tends to be impossible to substantiate.

In September 2017, Werrell and Femia (2017) commented on Selby et al. (2017) on *climateandsecurity.org*. They welcomed the study and identification of knowledge gaps but warned against underestimating links. They argued that Selby et al. did not refute the role of climate, but "muddied the waters" by (1) mischaracterizing science by arguing, against evidence, that drying prior to conflict was not due to climate change; (2) conflating causality and contribution by mispresenting advocates of climate-Syria link as identifying climate as primary cause; and (3) underestimating the human toll by selectively choosing migration data for an unrepresentative period.

Selby et al. (2017(b)) responded to Gleick and Kelley et al. in *Political Geography*, describing the claims of climate being an unspecified contributory factor as near-meaningless and unfalsifiable. They explained that they aimed to understand the quality of evidence for a climate-Syria link, which they considered weak for migration, grievances, and drought-climate links. They emphasized that Kelley et al.'s probabilistic results imply that the drought could have occurred without climate change. Finally, Selby et al. acknowledged that climate change will affect conflicts, but there are questions about where, when and how; and that unnuanced discourses risk fueling further climate skepticism.

Ide's (2018) article in *Current Climate Change Reports* synthesized the main studies of the climate-Syria debate, and highlighted perspectives to reduce polarization of the debate. He argued that the focus on high-profile cases leads to biases and reduces the quality of advice, and that polarization prevents complementarities between researchers and tends to favor extreme positions. Ide advocated more theoretically informed quantitative studies, mediated discussions between scholars with diverse positions, and stronger mixed methods approaches.

On 14 June 2019, Farhana Yamin, international environmental lawyer and Extinction Rebellion activist, published "This Is the Only Way to Tackle the Climate Emergency" in *Time* magazine. Yamin describes the familiar argument that in 2006–2011, Syria saw the worst drought and crop failures in its history, driving two to three million people to poverty and displacement and amplifying the factors that led to the war. She also stated that "something similar is occurring in Yemen".

Finally, in February 2022, the IPCC considered the evidence for climate-Syria links in its AR6 (2022). It reviewed 20 peer-reviewed articles, and concluded that while the 2006–2010 drought and the agricultural losses were linked with climate change, current evidence does not enable considering what might have happened in Syria in a non-climate changed world. The IPCC found that evidence is insufficient to show a causal link between the war and climate change, and that the 2011 uprisings would most likely have happened without the drought.

Linking climate change and Syria: controversies, translations and knowledge

The previous sections provided a brief history of the Syrian civil war and a chronological summary of the debate on the role of climate change in the conflict. As for Darfur, the next step is to identify the key controversies emerging from the contributions, followed by sections that describe the translations and knowledge resources mobilized by the actors.

The chronology illustrates how efforts to associate climate and the war involve recurring statements about the link between the two. However, and as seen in the case of Darfur, the efforts to translate climate change and the Syrian civil war together invite dissenting translations from others, and the statements evolve into controversies. Table 6.1 illustrates the initial and dissenting translations to associate climate change and the Syrian civil war.

These are the five main controversies within which climate and Syria are associated or disassociated. The next sections will describe each of these controversies, thereby highlighting in detail the translations that occur, both in terms of statements

Table 6.1 Main controversies about the climate-Syria link

Initial translation	Dissenting translation
Prior to conflict outbreak, Syria suffered its worst drought ever around 2006–2010.	The drought was less severe than claimed.
The drought was part of a long-term trend of anthropogenic climate change.	Drying trends are not obvious; attribution to climate is questionable; natural variability is high.
The rural crisis led to massive migration of population toward cities.	Migration was much less significant than claimed, and had other causes.
The influx of migrants destabilized Syria's already fragile cities (especially Dara'a), leading to an escalation.	Syrian protesters were about repression, not climate or drought.
Climate change facilitated the rise of ISIS in Syria (as well as Iraq).	No climate-ISIS link; such linking distracts from "real" national security issues.

about the climate-conflict links and in terms of knowledge resources assembled to try to settle the controversies.

Prior to conflict outbreak, Syria suffered its worst drought ever

Werrell and Femia (2012), were the first to describe that in 2006–2011, up to 60 per cent of Syria experienced the worst drought and crop failure since the beginnings of agricultural civilization. Similarly, Mohtadi (2012) elaborated that the 2006–2010 drought lasted four seasons, in comparison to usual one or two, and that precipitation was lower than during any previous drought. Mills (2012) described the drought as the longest in a century, and Friedman (2013) highlighted that the "drought was the worst in Syria's modern history and happened in the four years just before the revolution". In addition, the 2014 report by CNA considered that the conflict was preceded by five years of devastating droughts, Gleick described how Syria had a "multiseason, multiyear period of extreme drought" (2014, 332) and Kelley et al. (2015) referred to the worst drought on record starting in the winter of 06/07.

However, others questioned the severity of the drought. De Châtel (2014) argued that drought is a normal part of Syrian climate, and noted that while rains were poor in some areas in 08/09, they recovered elsewhere. Selby et al. (2017(a)) examined the claims about drought severity by Werrell and Femia (2012), Gleick (2014) and Kelley et al. (2015), highlighting that they did not provide rainfall data for Syria specifically, but used data for the broader Fertile Crescent. Selby et al. also noted that claims about the duration of the rainfall shock vary between authors and within the contributions (including 2007–2012, 2007–2010, 2005–2008, 2005–2010, and "anywhere up to six years") (Selby et al. 2017(a), 234).

In response to the criticisms, Kelley et al. argued that Selby et al.'s own data confirms the winters of 2006–2009 as the driest three years on record, and a very dry winter in 2010–2011 (2017, 245). The research on drought severity was summarized by the IPCC in 2022, with the conclusion that, in 2006–2010, the Fertile Crescent did have its worst drought on record, compounding by a consistent drying in the past 50 years.

Figure 6.1 provides an overview of the actors, translations and knowledge involved in the controversy about drought severity. Werrell and Femia (2012) cited an article by Gary Nabhan on *grist.org* – an online environmental news magazine – on the plight of Middle Eastern pepper farmers (Nabhan 2010):

> As one expert puts it, this may be the worst long-term drought and most severe set of crop failures since agricultural civilizations began in the Fertile Crescent many millennia ago.
>
> (Werrell and Femia 2012)

The identity of the expert is unclear, but she/he is also cited by Friedman (2012) and Gleick (2014). Friedman also travelled to Turkey and northeastern Syria to learn about the "Jafaf", or the drought. In Sanliurfa, he interviewed Faten, a Syrian

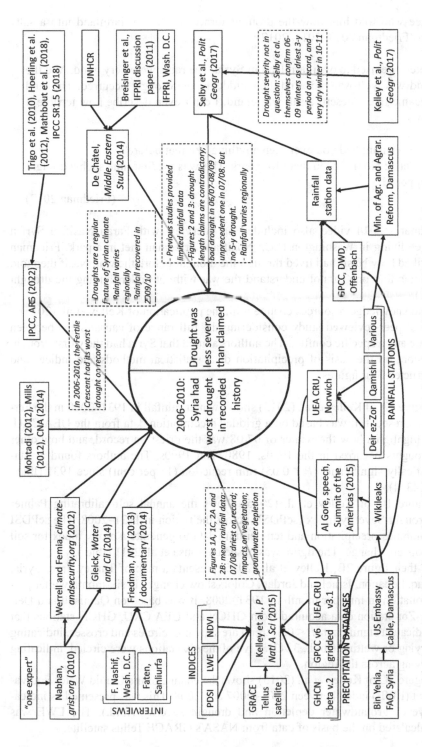

Figure 6.1 Main actors, translations and knowledge resources related to severity of Syria's pre-war drought

refugee who told him how the drought turned the family farmland into a salty desert (Friedman 2013, 2014, 2017):

> Faten: we used to own farmland in Syria. The rains were very good. And the land was well-watered. And then suddenly, the drought occurred. The land became like a desert, a salt wasteland. I can't even describe how terrible it was.
>
> . . .
>
> Friedman: had you ever seen anything like that before?
> Faten: No. I'd been there for years. This was the first time such a drought happened.
>
> (Friedman 2017)

Friedman's 2014 video also included an interview with Farah Nasif, a Syrian refugee living in Washington D.C.: Nasif and Friedman met in a park. Friedman described how Nasif had lived through the drought before the war. Nasif then told Friedman that one cannot understand the war without understanding the drought (Ibid.).

The knowledge resources change with the publication of Kelley et al. (2015) – the first peer-reviewed study considering the full range of causal steps between climate and a specific conflict. The authors argued that Syria had the worst drought on record on the basis of precipitation data, statistical methods and indices and constructed four figures:

- Figure 1A of Kelley et al. (2015) shows winter rainfall in 1931–2008 in the Fertile Crescent. It was based on a gridded precipitation data from the UEA CRU. It highlights how the winter of 07/08 was the driest on record, and how other droughts occurred in the 1950s, 1980s and 1990s. The authors found a "statistically significant" ($P < 0.05$) rain reduction (13 per cent) since 1931 (Ibid., 3243–4).
- Figure 1C of Kelley et al. (2015) shows the annual self-calibrating Palmer Drought Severity Index (scPDSI) across the region in 1931–2008. The scPDSI combines precipitation and temperature data to generate a proxy index for soil moisture change. The figure was based on Sousa et al. 2011 (Ibid.).
- In their figure 2B, Kelley et al. (2015) presents a map of Turkey, Iran, Syria, Iraq, Lebanon, Israel and Jordan. Colors show changes in six-month winter precipitation in mm per month in 1931–2008. It was based on Qamishli and Deir Er-Zor station data acquired from GHCN and UEA CRU. GHCN stations that indicate a significant ($P < 0.1$) trend are shown as circles and crosses, indicating drying or wetting (Ibid., 3243). Five of the six stations were circled, indicating drying across the region.
- Figure 2C of Kelley et al. (2015) shows the change of the Liquid Water Equivalent (LWE) index between 2001–2007 and 2008. This was presented as indicative of groundwater depletion and drought severity (Ibid.). The LWE was calculated on the basis of data from NASA's GRACE Tellus satellite.

In the critical contributions, De Châtel's questioning of drought severity drew on a 2011 paper by Breisinger et al., based on which she described how in 1961–2009, Syria had 25 years of drought, each lasting 4 ½ years, and one 10-year drought in the 1970s. She then highlighted, on the basis of a United Nations High Commissioner for Refugees (UNHCR) report (2011), that poor rains continued in 08/09 in Deir ez-Zor, Hassakeh and Raqqa but recovered elsewhere, and that rainfall recovered in 09/10 (though it remained irregular in the northeast).

Selby et al. (2017(a)) interrogation involved reanalyzing the data used by Kelley et al. (2015). They received data for five Syrian stations from Mr. Lister at CRU, finding that the 07/08 winter was the driest on record (precipitation 35 per cent below 1961–1990), and that 2006–2009 was the driest three-year period (precipitation 22 per cent below long-term). They observed that similar patterns are confirmed by data from Qamishli and Deir ez-Zor. However, Selby et al. also considered that the deviation in 2006–2009 was mainly due to the extremely dry winter of 07/08, that 06/07 and 08/09 were not exceptionally dry, and that in 09/10, rainfall was close to average. The authors illustrated this with three figures:

- Figure 2 of Selby et al. (2017(a), 236), which is based on the data from Mr. Lister, reflects the deviation of winter rainfall in 1948–2010 from 1961–1990 at stations in Turkey (Qamishli and Diyarbakir) and Syria (Deir ez-Zor). The figure shows a major dip in 07/08, while rains before and after that were close to average.
- Selby et al.'s (2017(a)) figure 3 (Ibid.), also based on Mr. Lister's data, indicates similar deviations for Aleppo, Deir ez-Zor, Diyarbakir, Qamishli and Siirt: a dip in 2007, dry conditions in 2006 and 2008, and near-average in 2009. Selby et al. also sought to corroborate the data with the Global Precipitation Climatology Center (GPCC) of the Deutscher Wetterdienst in Offenbach, Germany, captured in gridded datasets by Becker et al. 2013, which integrated measurements for 85,000 stations worldwide for 1901–2012 with grid resolutions between 0.5 and 2.5 degrees.
- Based on data from 15 Syrian stations from the Syrian Ministry of Agriculture and Agrarian Reform (MAAR), the authors presented a map (Ibid., 237) showing differences in rainfall in winters of 06/07–08/09 in comparison to 1982/83–09/10 at different Syrian weather stations. It indicates that rainfall was 36 to 25 per cent below average in Hasakah, Al-Rakka, Deir ez-Zor and Palmyra; 25 to 0 per cent below average at Qamishli, El-Bab, Idleb, Lattakia, Hama, Tartous and Dara'a; and 0 to 18 per cent above average in Aleppo, Safita, Homs and Damascus. Based on this, the authors described that while the northeast was dry, Aleppo, Damascus and Homs had above average rains, and Dara'a (which advocates of climate-Syria links had described as severely hit by drought), had close to average rainfall (Ibid., 234).

Based on these materials and presentations, the authors concluded that northeastern Syria had a bad drought in 2006–2009, and an unprecedented one in 08/09, but it was not a five-year drought as claimed by the previous contributions, and there were several regional variations (Ibid., 237).

The IPCC's 2022 observations on Syrian drought severity were based on Trigo et al.'s 2010; Hoerling et al.'s 2012, as well the IPCC's 2018 report *Global Warming of 1.5°C*, which included Box 3.2 that synthesized studies related to Mediterranean drought.

The drought was part of a long-term trend of anthropogenic climate change

However, the mere existence of a drought would not demonstrate the workings of anthropogenic climate change. Thus, efforts were made to link the drought with climate change by arguing that the 2006–2010 drought was climate-related, and by mobilizing knowledge resources in support of those arguments.

Werrell and Femia (2012), Friedman (2012), Mohtadi (2012), and Mills (2012) were the first to argue that evidence links climate and the drought, and that the drought cannot be explained by natural variability. Similar views were expressed by Gleick (2014), who described 20th-century drying trends in the Mediterranean as well as patterns of decreasing rain and increasing evaporation. Friedman's segment in *Years of Living Dangerously* (2014, 2017) continued this line of argument, as did adelphi's 2015 video. Kelley et al.'s 2015 study sought to exclude natural causes of the drying and warming trends, presented findings of climate models considered consistent with increased GHG emissions, and concluded that the 2006–2010 drought was more than twice as likely due to climate change.

Criticisms of climate-drought links were again made by De Châtel and Selby et al. De Châtel described the "extent to which climate change exacerbated the situation" as "debatable" (2014, 522). Selby et al. (2017(a)) questioned whether there was a long-term drying trend in the region – if there was none, then a link to increasing GHG emissions would be unlikely. In their criticisms, they highlighted:

- Only 5 of the 25 rainfall stations analyzed by Kelley et al. showed linear drying trends, those five were "of marginal significance", and only one was in Syria.
- Hoerling et al.'s analysis of Mediterranean drought considered no stations from Syria.
- While Kelley et al. and Hoerling et al. observed dry periods in 1999–2008 and 1985–1995, respectively, the linear trend for the 20th century has been weak, and there is high inter-annual and inter-decadal variability in rainfall.
- The linear modeling of rainfall applied by Kelley et al. can be misleading due to the high inter-annual and -decadal variability (Ibid., 241–2).

On this basis, Selby et al. considered that there has been no progressive drying in the Fertile Crescent, and thus no basis for linking climate change and the drought (Ibid., 235).

Their second point was to question Kelley et al.'s method to attribute the drought to climate change with the following observations:

- Attribution methods, data and models are still at initial stages.
- Links between human influence and regional rainfall are not fully understood.

- Kelley et al.'s attribution method was to: (1) identify a long-term drying trend; (2) estimate the likelihood of a drought given this trend; and (3) compare the trend with model simulations. But since there was no long-term trend, the other assumptions are invalid.
- The models used by Kelley et al. involve biases of up to 40 per cent.
- Even if climate change increases droughts, implicating it for a specific drought confuses a probabilistic claim with a deterministic one (Ibid., 242).

Selby et al. concluded that there is no clear evidence that climate influenced the drought (Ibid. 235–7). The points by Selby et al. elicited the following reactions:

- Selby et al. provided no evidence for their claim that the dry period in the last 10–25 years was due to natural variability (Kelley et al. 2017, 245).
- Other work with different methods has shown that long-term drying trends agree with human-induced climate change, climate models, and a "well-established theory of subtropical winter circulation and climate change" (Ibid.).
- The period of 1998–2012 was the driest 15-year period in the last 900 years – this confirms that recent drying is outside natural variability (Ibid., 246; Werrell and Femia 2017).
- Literature supports the notion that the trend is likely due to a changing climate, and downplaying the role of climate creates risks of inadequate preparation and closes "avenues of conflict resolution and peacebuilding" through water sector cooperation (Werrell and Femia 2017).

The IPCC's AR6 also provided specific observations about links between climate change and the drought, considering that the drought is attributable to GHG emissions (2022, 16–24).

Figure 6.2 illustrates the actors, knowledge resources and key translations in relation to climate-drought links. In early 2012, the *Journal of Climate* published "On the Increased Frequency of Mediterranean Drought" by researchers at the NOAA Earth System Research Laboratory in Boulder, Colorado (Hoerling et al. 2012). The article became a reference point for climate-Syria links. It was first cited by Werrell and Femia (2012) from *thinkprogress.org* by Joseph Romm – a physicist and environmental blogger. Romm cited a NOAA press release of the study, describing it as a "bombshell". Werrell and Femia noted:

A NOAA study . . . found strong and observable evidence that the recent prolonged period of drought in the Mediterranean littoral and the Middle East is linked to climate change. On top of this, the study also found worrying agreement between observed climate impacts, and future projections from climate models.

(Werrell and Femia 2012)

Similarly, Friedman (2012) cited Romm's entry, highlighting that winter droughts in the Middle East are increasing partly due to climate change. Friedman (2012)

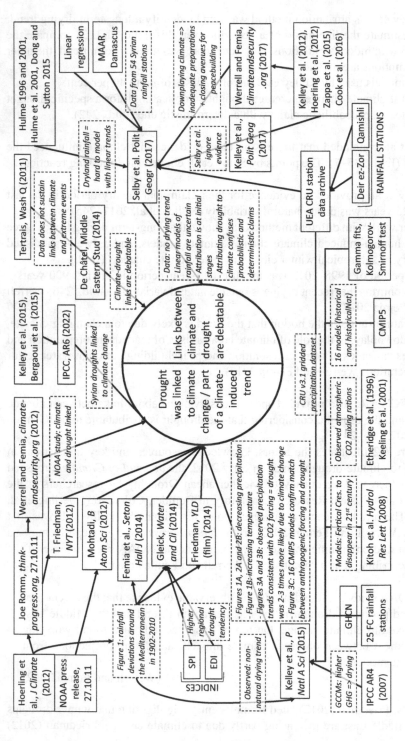

Figure 6.2 Main actors, translations and epistemic resources related to controversies about links between climate change and Syria's prewar drought

and Mohtadi (2012) quoted Hoerling as stating that the "magnitude and frequency of the drying . . . is too great to be explained by natural variability". Mohtadi, while highlighting that one drought cannot be a direct result of climate change, also described on the basis of the NOAA press release, that "climate change from greenhouse gases explained roughly half the increased dryness of 1902–2010".

In 2014, Peter Gleick cited Hoerling et al. and the press release, including that the drying cannot be explained by natural variability, and that climate change is reducing winter rain and increasing evaporation, leading to more droughts (2014, 337). He also replicated two figures from Hoerling et al.: one showing rainfall anomalies around the Mediterranean as deviations in millimeters in the winter period from 1902 to 2010, with increasing drops (identified as red bars) in 1987–1994, 1997–2002 and 2005–2008 (Ibid., 336); and a map of the Mediterranean showing the regional distribution of rainfall anomalies. Syria, Jordan, the Adriatic coast and parts of Portugal are bright red, indicating significant reductions. Gleick explained that the "reds and oranges highlight the areas around the Mediterranean that experienced significantly drier winters during 1971–2010 in comparison to 1902–2010" (Ibid.). Gleick also described that the Mediterranean drought had changed from historical norms. He cited a study by Mathbout and Skaf – researchers in Latakia, Syria – who used the standard precipitation index (SPI) and effective drought index (EDI) to identify an increased regional drought tendency (Ibid., 336). He also cited studies that identified a "statistically significant increase" in evaporative water demand in 1988–2006 in the region, driven by higher sea level surface temperatures (Ibid., 337).

Femia et al. also described how Hoerling et al. found evidence linking the drought to climate change, and cited that the droughts were "manifestations of an extended period of climate change-related winter drying beginning in the 1970s" (2014, 77). They also reproduced Hoerling et al.'s two graphs, though with modified headings: Hoerling et al. described the map of the Mediterranean as follows: "The observed change in cold season precipitation for the period 1971–2010 minus 1902–70. Anomalies (mm) are relative to the 1902–2010 period. Solid curve is the smoothed precipitation time series using a nine-point Gaussian filter. Data are from the GPCC" (2012, 2147). Femia et al. retitled the map as follows: "Lands around the Mediterranean that experienced significantly drier winters from 1971–2010 than the comparison period from 1902–2010" (2014, 78). Similarly, Femia et al. replicated Hoerling et al.'s figure on chronological rainfall deviations, originally titled "Observed time series of Mediterranean (308–458N; 108W–408E) cold season (November–April) precipitation for the period 1902–2010" (Hoerling et al. 2012, 2147), relabeling it to "Winter precipitation trends in the Mediterranean region for the period 1902–2010" (Ibid., 78).

In parallel to these two publications, the link between climate change and the prewar drought was also considered in Friedman's segment in *Years of Living Dangerously* (2014, 2017). Friedman stated that the drought was part of a trend: according to NOAA, climate change has dried the Mediterranean region, worsening droughts, with Syria "at the epicenter". The video shows Hoerling et al.'s figure and includes a visually more dramatic version of it. The material is also replicated by the YCC video (2015).

Another video contribution referencing Hoerling et al. was published by Adelphi (2015). The video presented a conceptual model comprising a flow diagram that illustrated the steps from climatic change to conflict outbreak. It described that the Mediterranean littoral has experienced considerable drying, with 10 out of 12 driest winters since 1902 occurring in last 20 years. It added, with reference to Hoerling et al., that natural variability cannot explain this. Instead, anthropogenic climate change, aerosol forcing and sea surface temperature increase are identified as "major contributing factors".

So far the drought-climate link was based on individual statements, replication of graphs, and video materials based on Hoerling et al. However, Hoerling et al. themselves remained silent about the origins of the Syrian conflict. Rather, others translated their findings about drought to serve as evidence for climate-Syria links. This changed in 2015 with Kelley et al.'s study on climate-drought links.

Kelley et al. conducted complex translations with multiple knowledge resources and representations to demonstrate climate-drought connections. They linked the two by illustrating that (a) the drought was part of a longer-term drying trend illustrated by declining rainfall and temperature and (b) the observed trend can only be explained by applying models that consider anthropogenic climate change. Kelley et al. observed that four of the worst regional droughts occurred in last 25 years. Citing Kelley et al. (2012(a), 2012(b)) and Hoerling et al. (2012), they stipulated that there had been an "externally forced winter drying trend" in last half of 20th century, "distinguishable from natural variability" (Kelley et al. 2015, 3243).

To see if the drought was strengthened by long-term trends, Kelley et al. first determined the long-term rainfall change. They did this with their Figure 1A which shows the chronology of winter rainfall in 1931–2008 based on a gridded dataset from the UEA CRU. They also observed in their Figure 2B (map of Syria and surrounding countries) that the linear precipitation trends of GHCN stations corroborate the drying. An additional step was to plot rainfall data from UEA CRU 3.1 gridded dataset on a similar map, presented as Figure 2A, which they described as "similar to the climatological rainfall pattern" (Ibid., 3242). In addition, Kelley et al. presented their Figure 1B, which shows the measurements for annual mean surface temperature across the Fertile Crescent in 1931–2008, again based on CRU's data, concluding that there has been a "statistically significant ($P < 0.01$)" increase in temperature, with much of it in the last 20 years (indicated by red shading). They also illustrated, using Figure S2, that this trend was important during the summer, leading to higher evaporation, which, in combination with low precipitation, had significant impacts on winter crops, which depend on soil moisture and winter precipitation (Ibid., 3243). In addition, they observed, on the basis of their Figure 1C, how the scPDSI confirms this long-term trend. The authors considered that the 100-year and "statistically significant" trends suggest anthropogenic influence. They acknowledged that natural variability cannot be fully ruled out, but emphasized that the trends and increase in multiyear droughts and temperature are consistent with anthropogenic climate forcing (Ibid., 3244).

Kelley et al. then estimated the impact of the drying trend on drought likelihood. First, they separated the precipitation trend from natural variability "by regressing the running 3-y mean of observed (CRU) 6-month winter precipitation onto the running 3-y mean of observed annual global atmospheric CO_2 mixing ratios from 1901–2008" (Ibid.). In other words, they compared the difference between observed precipitation and CO_2 trends. They captured this in their Figure 3A, which chronologically shows the precipitation trend, the CO_2 trend, and their differences. They then constructed a frequency distribution of the timeseries with CO_2 and without CO_2, captured in their Figure 3B, applying gamma fits and the Kolmogorov-Smirnoff test. From this they concluded that "natural variability and CO_2 forcing are 2 to 3 times more likely to produce the most severe 3-year droughts than natural variability along" (Ibid.).

Kelley et al. drew further support from model simulations. Based on IPCC 2007, they observed that global circulation models agree that the region will become drier with higher GHG levels, and, based on a study by Kitoh et al. 2008, that high-resolution models indicate that the Fertile Crescent is likely to disappear by the end of the century due to climate change (Ibid., 3242). They also cited 16 models of the Coupled Model Intercomparison Project phase 5 (CMIP5), comparing model runs that consider all external factors, including GHG concentrations (called historical runs), with runs that consider only natural factors (called historicalNat runs). In their Figure 3C, they compared the observed and modeled trends, observing that severe droughts occur "less than half as often under natural forcings" (Ibid., 3244), thus supporting the attribution of Syria's pre-war drought to anthropogenic climate change.

De Châtel's (2014) initial criticism of climate-drought links cited Tertrais' (2011) observation that assumptions about warming leading to more extreme events is based on modeling, and that data does not sustain such a hypothesis. Selby et al. (2017(a)) questioned two aspects of the arguments for climate-drought links: (a) that there was a drying trend, and (b) that methods for attributing the trend to climate change are robust. Their analysis of the linear trends was based on rainfall data from

- rainfall stations in Qamishli and Deir-ez-Zor, drawn from UAE CRU's station data archive;[1] and
- the Syrian Ministry of Agriculture and Agrarian Reform (MAAR), whose data covered 54 measurement stations for 1982/83–2009/10.

To see whether there was a linear drying trend, the authors used linear regression on the data from Qamishli and Deir-ez-Zor, observing that

although rainfall declined on average through their 60 (Qamishli) and 65 (Deir ez-Zor) year periods, this decline was nowhere near significant for the former station (p = 0.72) and only close to significant at the 5% level for the latter (p = 0.06). In line with our argument above, moreover, the trends over the twentieth century (up to 1998/99) are even less significant (p = 0.30 and p = 0.34 respectively).

(Selby et al. 2017(a), 242)

In addition to rainfall data, Selby et al. cited articles by Hulme (1996) and Hulme (2001) to stipulate that dryland rainfall involves high variabilities and is difficult to model linearly (2017(a), 235) and highlighted how Dong and Sutton (2015) and Hulme (2001) have shown that linear modeling of dryland rainfall is misleading. Their discussion on limitations of attribution cited Bellprat and Doblas-Reyer (2016) and Hulme (2014). In terms of climate models, Selby et al. highlighted He and Soden (2017), which questioned "the reliability of the sub-tropical precipitation decline thesis, suggesting that in response to human influences on the atmosphere rainfall declines in the sub-tropics are more likely over ocean areas (e.g. over the Mediterranean Sea) than over land (e.g. northeast Syria)" (Ibid., 242).

The responses by Kelley et al. (2017) and Werrell and Femia (2017) to the scrutiny of climate-drought links were based on the following materials:

- To further demonstrate climate-drought links, and illustrate that Selby et al. ignored a large body of literature, Kelley et al. cited in particular:

  - Kelley et al. (2012) in *Climate Dynamics* and Kelley et al. (2012) in *Geophysical Research Letters*, which found that only anthropogenic change can explain the drying;
  - Hoerling et al. (2012), Zappa et al. (2015) in *Climate Dynamics*, which observed human-driven winter drying in the Fertile Crescent;
  - Zappa et al. (2015) in *Environmental Research Letters*, who considered that newest models actually underestimate Mediterranean drying;
  - Cook et al. (2015) in *Science Advances* and Cook et al. (2016) in *Journal of Geophysical Research*, which described tree-ring datasets that indicate that 1998–2012 was the driest period in the Levant in the last 900 years;

- Werrell and Femia also referenced the above studies, as well as the tree ring studies by Cook et al. from Kelley et al. (2017).

IPCC's 2022 conclusion that the drought was attributable to climate change was based on Kelley et al. (2015), from which the IPCC observed that the "magnitude of the multiyear drought is estimated to have become two to three times more likely as a result of increased CO_2 forcing" (2022, 16–23). The IPCC also cited a study by Bergaoui (2015), which observed that a later drought in Syria (2013–2014) was 45 per cent more likely due to anthropogenic climate change.

The drought led to massive migration of population toward cities

Regardless of whether the drought was climate-related, most contributions considered that it forced rural inhabitants to move from the northeastern governorates to Syria's cities. Werrell and Femia (2012) spoke of a "massive exodus". On NPR, Nayan Chanda, editor-in-chief of *YaleGlobal Online* magazine, described this as the "largest internal migration in the Middle East" (NPR 2013). Faten, a Syrian refugee living in Sanliufra, Türkyie, told Friedman (2013) how her family moved due to the drought. Videos by YCC (2015) and Adelphi (2015) elaborated how the drought brought a diaspora from farms to overcrowded cities. Damascus, Aleppo, Dara'a. Hama, Homs and Deir-ez-Zour were highlighted as migration destinations (Mohtadi 2012; de Châtel 2014; Gleick 2014).

Many contributions estimated the scale of drought-induced migration, in both absolute numbers and proportional amounts. Werrell and Femia (2012) and Femia et al. (2014, 76) stated that just around Aleppo, 200,000 people were forced to migrate. De Châtel referred to 300,000 people (2014, 527), the CNA to "millions" (2014, 14), and Wendle (2016, 52) and Friedman (2014, 2017) to more than 1 million. The estimated numbers of families included 50,000 (Femia et al. 2014, 76), 65,000 (de Châtel 2014) and 300,000 (Kelley et al. 2017). In terms of proportional numbers, Friedman (2013), spoke of migration of "half the population" between Tigris and Euphrates, and De Châtel described that 60–70 per cent of villages in Hassakeh and Deir ez-Zor were deserted (2014, 527).

However, one number gained special traction: that one and a half million people were displaced by the drought. Many contributions adopted this figure, and it circulated in research papers (Femia et al. 2014; Gleick 2014; Kelley et al. 2015), media (Mohtadi 2012; NPR 2013; Gray 2015; Zastrow 2015, Wendle 2016), NGO contributions (YCC 2015) and speeches (Gore 2015; Kerry 2015). For example:

It is estimated that the Syrian drought has displaced more than one and a half million.

(Mohtadi 2012)

More than 1.5 million people . . . moved from rural land to cities and camps on the outskirts of Syria's major cities.

(Gleick 2014, 334)

Estimates of the number of people internally displaced by the drought are as high as 1.5 million.

(Kelley et al. 2015, 3242)

As many as 1.5 million people migrated from Syria's farms to its cities, intensifying the political unrest that was just beginning to roil and boil in the region.

(Kerry 2015)

While all agreed that some migration was triggered by the drought, the numbers, in particular the one and a half million, evolved into a controversy. Fröhlich, for example, observed that "exact numbers are scant" (2016, 40). Selby et al. (2017(a)) systematically considered the sources of the number. They highlighted that the one and a half million figure comes from limited sources, possibly means the number of people affected – not displaced, and is inconsistent with other estimates made at the same time. Selby et al. considered 40,000–60,000 families plausible.

A second point of criticism was the importance of the drought as a driver of migration vis-a-vis other factors. Most contributions, including those advocating a climate-Syria link, acknowledged that other factors played a role (Femia et al. 2014; Gleick 2014; Wendle 2016). De Châtel described how cuts in fuel and fertilized subsidies drove impoverished farmers to cities (2014), and Fröhlich considered that migration decisions usually involve many economic, political, demographic and environmental factors, and are often mediated by institutions, structures, networks, and patterns of chain migration (2016, 38).

Selby et al., also argued that the inattention of Gleick, Kelley et al. and Werrell and Femia to changing economic factors led them to overstate the impacts of the drought (2017(a), 238). They described that while the three recognized the multiple causes of migration, they did not account for the relative role of those causes. Selby et al. considered that previous contributions overestimated the role of drought-induced migration (Ibid., 239), and described Kelley et al. as arguing that drought migration was central to population pressures because it accounted for about half of the urban growth in 2003–2010, and that rapid demographic change encourages instability. They further disputed the centrality and scope of drought migration by identifying six other drivers of migration and migratory patterns that existed within Syria in 2003–2010 and led to changes in urban populations.

In response to the criticisms, Kelley et al. replied to Selby et al.:

> The severe drought triggered an agricultural collapse and an internal displacement of entire rural farm families utterly unlike the usual seasonal labor migrations or the rural-to-urban migration from prior years.
>
> (Kelley et al. 2017, 246)

Kelley et al. also argued that Selby et al. gave little evidence that excess drought migration was only a small proportion of total displaced (Ibid.). In their response to Selby et al., Werrell and Femia highlighted:

> The drought was one of a number of other environmental, economic and governance factors – including natural resource mismanagement by the Assad regime – contributing to the mass displacement of a significant number of Syrians, and thus potentially increasing the "likelihood" of conflict – a probabilistic, not a causal, claim.
>
> (Werrell and Femia 2017)

Again, the IPCC reviewed what existing scientific publications say about drought-related migration, highlighting that the drought aggravated water insecurities (2022, 4–53), devastated agricultural production, and forced people to leave their homes (Ibid., 16–23). The report emphasized that the role of drought in increasing migration remains contested, as do the levels of migration (ranging from 40–60,000 families to one and a half million people), and that migration numbers need to be seen in context. The IPCC concluded that while agricultural impacts can be linked to the drought and thus partly to GHG emissions, its impact relative to other drivers has not been adequately quantified (Ibid., 16–24).

What were the origins of the migration estimates? Figure 6.3 shows the actors, knowledge resources and translations. Werrell and Femia 2012 cited the *Global Assessment Report on Disaster Risk Reduction* to state that nearly 75 per cent of vulnerable Syrians "suffered total crop failure" and one million were left food insecure. They also referenced Worth's 2010 *NYT* article, which, in turn, cited an UNOCHA survey that found that northeastern farmers lost 85 per cent of livestock, and 1.3 million were affected. Friedman (2012) cited Werrell and Femia on the scope affected land, crop and livestock losses, affected people, and those who lost their livelihoods.

For specific migration numbers, UNOCHA's *Syria Drought Response Plan 2009–2010: Mid-Term Review* was the basis for De Châtel's estimates of 300,000 people, 65,000 families, and 60–70 per cent of villages in Hassakeh and Deir-ez Zor. This was also cited by Ababsa 2015 to claim that 300,000 families migrated, a figure later referenced by Kelley et al. 2017. The number of 50,000 families came from Worth's 2010 article, who cited a "UN estimate" for this number. The migration of 200,000 around Aleppo was based on Nabhan (2010), and Friedman's estimate of "half the population" between the Tigris and Euphrates was drawn from a statement by Samir Aita, a Syrian economist. Others, mainly CNA (2014), Wendle (2016), Friedman (2017), did not identify any sources for the figure of "millions".

Given the centrality of the one and a half million figure, it is worth looking at its exact origins. The number first appears in a 2009 report by IRIN, in a quote from Mohamad Alloush, Director of Environmental Department of the State Planning Commission of Syria (SPC), who is quoted as saying:

> The drought also forced 250,000–300,000 families (at least 1.25–1.5 million people) to leave their villages and they are now concentrated in the suburbs of Damascus and other cities like Aleppo and Da'ra.
>
> (IRIN 2009)

Originally, Mohtadi (2012) and Kelley et al. 2015 used this source. Then, Femia et al. 2014; Gray 2015 cited it from Mohtadi, and Zastrow (2015) and Wendle (2016) from Kelley et al. YCC (2015) integrated it from CNN footage about Syria. Many others (for instance NPR 2013; Gleick 2014; Gore 2015; Kerry 2015) cited the number without specifying its sources.

While IRIN published the quote, different Syrian government estimates were already circulating. A 2010 article by Mahmoud Solh, director general of the International Center for Agricultural Research in the Dry Areas (ICARDA) in Syria, cited Syrian government and UN sources which spoke of the displacement of 40,000–60,000 families, or 200,00–300,000 people. A report by Massoud (2010) highlighted three government sources: (1) MAAR (40,000–60,000 families); (2) Baath Party conference in Qamishli (200,000 people); and (3) "other estimates" (30,000 families from Qamishli and 50,000 from al-Shadadi). Kelley et al. cited all three, but highlighted the one and a half million estimate.

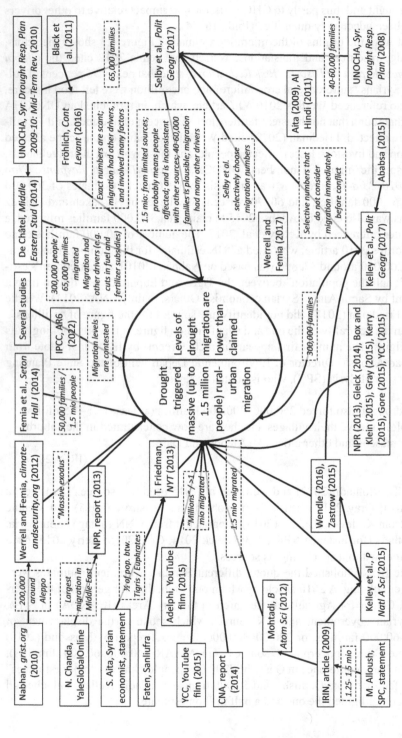

Figure 6.3 Main actors, translations and knowledge resources related to the controversy about levels of migration caused by Syria's pre-war drought

In reaction to migration numbers, in particular the one and a half million figure, Selby et al. acknowledged that the drought had migration impacts, as illustrated by sources such as the UN, Syrian government, US diplomatic channels, media, interviews and UNOHCA drought appeals (2017(a), 237). However, they asked whether the numbers were plausible. Referring to IRIN, they highlighted that the one and a half million estimate appears to be based only on the statement by Mr. Alloush, and considered that the number possibly referred to people affected, not displaced. They highlighted that the UNOCHA *Syria drought response plan* (August 2009) stated that one and one third million people were affected (Ibid.), and listed other estimates:

- A UNOCHA September 2008 drought appeal stated that migration was 20–30 per cent higher due to drought.
- UN estimates in the *Syria drought response plan* and a Joint UN Drought Assessment Mission referred to 40–60,000 families, including 36,000 (200–300,000 people) from Hasakah.
- UNOCHA's *Syria Drought Response Plan 2009–2010 Mid-Term Review* spoke of 65,000 families.
- A press release by Olivier de Schutter, the UN Special Rapporteur on the right to food, which stated that 600,000 had migrated. De Schutter presented this after his visit to Syria in 2010.[2]

Based on this, Selby et al. concluded that the most cited figure is 40,000–60,000 families (2017(a), 237).

As examples of other drivers of migration, Selby et al. highlighted the economic transformations during the drought, including privatization of state farms, liberalization, and removal of fuel and fertilizer subsidies,[3] emphasizing that these impacted agriculture and rural-to-urban migration, which was already high before the drought (about 135,000 people per year in 2000–2005) – based on al Hindi 2011. In addition, Selby et al. identified other drivers of migration:

- Syria's normal population growth of three per cent annually (based on Aïta 2009);
- One and a half million Iraqi refugees (based on Kelley et al. 2015);
- General rural-urban migration of about one million (extrapolated from al-Hindi 2011);
- Outmigration of approximately 900,000 (Aïta 2009);
- End of circular migration to Lebanon, possibly several hundred thousand (Chalcraft 2008);
- Excess migration from Northeast in 2008–2009, estimated at 40,000–60,000 families.

On this basis, Selby et al. considered that not all migration was due to drought, and thus previous studies must have overestimated the role of drought migration in population pressures.

For their response to Selby et al., Kelley et al. (2017) reiterated that the number of migrants is likely to be close to one and a half million, stating that Selby et al. had selectively cited Ababsa 2015, and did not adequately consider the "displacement that occurred in the 6–18 months directly preceding the uprising" (Kelley et al. 2017, 246). They then referenced Ababsa's estimates as follows: 160–220 villages abandoned (based on a report by International Institute for Sustainable Development) and 300,000 families driven to cities (from UNOCHA 2010) (Ababsa 2015, 199). However, Ababsa did not clearly identify which UNOCHA document is referred to. One possibility is the *Syria Drought Response Plan 2009–2010 Mid-Term Review* of February 2010. However, UNOCHA referred to 65,000, not 300,000, displaced families (UNOCHA 2010, 6). As additional evidence, Kelley et al. cited interviews conducted by Wendle in refugee camps in Greece, published in *Scientific American* (Wendle 2016), arguing that "all agreed that the drought instigated a mass migration" (2017, 243).

Similarly, Werrell and Femia (2017) considered that Selby et al. selectively chose sources and excluded later stages of the drought and reports of displacements in 2009–2010. They reiterated how Kelley et al. 2017 cited Ababsa's understanding of UNOCHA's 2010 report that 300,000 families were driven to cities by the drought. They also argued that one should not accept only official Syrian and UN estimates for one region and one year, and that collectively, journalistic, government UN and policy institute sources support that one and a half to two million people were displaced during the drought in 2006–2011.

The IPCC's review of the migration controversy cited in particular:

- Kelley et al.'s (2015) observations about the drought as a factor in aggravating water and agricultural insecurities;
- De Châtel (2014) and Werrell et al. (2015) on the drought's impacts on agricultural production;
- Kelley et al. (2015) and Gleick (2014) on migration effects of drought, and Selby et al. (2017(a)) and Gleick (2014) for migration levels (40–60,000 families and one and a half million people, respectively);
- Observations about differences in drought effects in Turkey, Iraq and Jordan from studies by Trigo et al. (2010) and Eklund and Thompson (2017);
- De Châtel (2014), to emphasize that migration numbers need to be seen in context;
- Several studies (Aïta 2009; De Châtel 2014; Selby 2019) to highlight that impacts vis-à-vis other drivers have not been quantified.

Migration destabilized Syria's fragile cities, leading to war

So far, this chapter has described how climate change and the Syrian conflict were linked by translations stipulating that prior to the conflict, Syria had the worst drought in its recorded history, that the drought was more likely due to climate change, and that it triggered mass migration to cities. Others questioned the drought severity, the evidence for climate-drought links, and the estimated levels of migration. The chapter also described the knowledge resources mobilized by actors to debate the controversies.

The final component of climate-Syria mechanism is the assumption that Syrian cities were destabilized by migration pressures or demands of migrants, to which the Assad regime responded brutally, triggering the war. Within this component, the contributions focused on three aspects: first, Syrian cities were already coping with pressures prior to the arrival of migrants. Second, that migration exacerbated those pressures into a breaking point. Third, that protests began in places that had been hit hard by the drought or migration, in particular the southern town of Dara'a.

Several contributions highlighted existing pressures on Syrian cities (Kelley et al. 2015; YCC 2015; Zastrow 2015). Mills (2012) described how the slums of Syrian cities, previously full of refugees, were now teeming with rural migrants. YCC (2015) integrated CNN footage with Christiane Amanpour explaining how drought migration happened when Syrian cities were under "enormous economic and social pressures".

Kelley et al. focused on population growth as a pressure factor, highlighting that the urban population of Syria had increased from 8.9 to 13.8 million in 2002–2010. They illustrate this with a line chart (their Figure 1D) that showed Syria's populating growth from 4 to 22 million between 1950 and 2010 (2015, 3242). A second factor, the presence of up to two million Iraqi and Palestinian refugees, was described by Mills 2012; Mohtadi 2012; Werrell and Femia 2012; Femia et al. 2014 and Kelley et al. 2015. According to Kelley et al., refugees constituted 20 per cent of urban population prior to the conflict (2015, 3242). A third factor, highlighted by Werrell and Femia 2012 and Femia et al. 2014, was that Syrian cities had water infrastructure deficiencies. The contributions also described other factors, such as environmental problems (Femia et al. 2014, 76), overcrowding, unemployment, corruption and inequality (Kelley et al. 2015, 3242).

Next aspect of this controversy was how the cities were strained by the aforementioned factors (Mills 2012; Mohtadi 2012; Werrell and Femia 2012; Gleick 2014; Femia et al. 2014; YCC 2015; Zastrow 2015). Gray summarized this as follows: migration caused poverty, unrest, and pressure on urban areas, which led to uprisings and war (2015). While acknowledging the difficulties of specifying the degree to which displacement and rural disaffection drove unrest (Femia et al. 2014, 79), several contributions described how pressures "strained" Syrian cities and contributed to social problems. The arguments also involved theoretical observations. Kelley et al. argued that, according to "conflict literature", rapid demographic change encourages instability, and that drought-induced migration exacerbated other factors such as unemployment, corruption and inequality (Kelley et al. 2015, 3242).

A third type of pressure was the notion of resource competition. For instance, Kelley et al. argued that the "population shock" from drought migration increased strain on resources (2015, 3242), and, consequently, the poor were forced to compete for employment and water (Werrell and Femia 2012; Femia et al. 2014, 79). As other areas of competition, adelphi's conceptual model specified how the exodus to cities triggered competition for housing, jobs and services (Adelphi 2015).

In addition to observations about destabilized cities, several contributions argued that unrest erupted in areas that were most affected by drought or were migration destinations (Mohtadi 2012; Gleick 2014). Others, including Werrell and Femia (2012), Femia et al. (2014), and Friedman (2014, 2017), described how disaffected rural communities were prominent in the protests. Farah Nasif, in an interview with Friedman in *Years of Living Dangerously*, stated that "most people in the revolution are from countryside".

The city of Dara'a, where early protests happened, was connected with migration by several contributions (Mills 2012; Werrell and Femia 2012; Friedman 2013; NPR 2013; Gleick 2014; Femia et al. 2014; Friedman 2014, 2017). NPR host Jacki Lyden described how "the spark that triggered the civil war was ignited in the city of Dara'a in February 2011" (NPR 2013). Werrell and Femia (2012) and Femia et al. described Dara'a as especially hard hit by the drought, and as a focal point for protests (2014, 79). Gleick considered that unrest began in Dara'a, which saw a large influx of farmers displaced by crop failures (2014, 335). He also elaborated how Dara'a, historically a bread basket of Syria, was crippled by drought, and that in "all centers of popular uprisings" lie narratives of lost livelihoods (Ibid., 336). Friedman (2013, 2014, 2017) noted how drought refugees joined protests in Dara'a in 2011.

In contrast, Zastrow (2015) considered that it is not possible to evaluate whether violence would have occurred without the drought, that data is scarce, and people tend to give inaccurate answers when interviewed. Fröhlich (2016) observed that there is no evidence of causality between climate migration and conflict and that migration can be an adaptation mechanism. She called for more nuanced approaches to considering social, demographic, political and economic drivers of migration. Along similar lines, Selby et al. argued that conflict literature does not support links between demographic change and conflict, that protesters were not concerned about drought, and that people from the northeast were not involved. In addition, they criticized the lack of individual testimonies in Kelley et al. (2015) and Gleick (2014), whose assumptions about the role of migrants in protests were based on quotes from one farmer and a leaked US Embassy cable.

The narratives of the situation in Dara'a also turned controversial. Selby et al. highlighted that the first protests happened in Damascus, then in Kurdish areas, and later in Dara'a (Selby et al. 2017(a), 239). They described contradictory evidence for the economic situation in Dara'a: some said poverty increased, while others said it decreased (Ibid.). De Châtel argued that media reports wrongly stated that Dara'a was severely hit by drought (2014, 524) and that precipitation in Dara'a was normal prior to the conflict (Selby et al. 2017(a), 240). The critical contributions also emphasized the central role of repression in the uprisings (Ibid.). De Châtel described that the protests in Dara'a began when 15 schoolchildren were arrested and tortured by security forces (2014). Similarly, Fröhlich considered the causes of protests:

Two women from different Dara'a clans had been arrested and abused by authorities. This was, as the story goes, followed by anti-regime graffiti which had been drawn by 15 schoolchildren in their defence. The children were subsequently arrested and tortured. Attempts to mediate their release were met by rejection and insult. The local security chief, Atef Najeeb, shockingly suggested "sending local women to conceive some new kids [to replace the arrested ones]" (Leenders 2013, 279). Consequently, the people of Dar'a rallied in protest and defiance, purportedly shouting 'to hell with you' to security forces who opened fire.

(Fröhlich 2016, 45–6)

In addition, Selby et al. identified political opportunity,[4] liberalization and subsidy removal as further causes of protests (2017(a), 240).

Fröhlich's study on the role of migrants in Dara'a protests (2016) asked if migrants could have contributed to the uprisings in Dara'a. She considered that collective action requires networks and trust, in particular in autocratic regimes like Syria. She then outlined three potential drivers of mobilization: opportunity, perception of threats to way of life and liberation. Fröhlich argued that specifically in Dara'a, the dense networks of the local population enabled mobilization in response to the threat from regime oppression, but the migrants, being an "out-group", did not share these networks or sense of community. She also highlighted that refugees had no drought-related grievances but were protesting against repression, and concluded that "climate migrants could not have been active protesters because of their lack of networks" (Ibid., 38).

The IPCC's AR6 also synthesized the peer-reviewed part of the debate about the role of migration in the outbreak of the war and in the protests in Dara's and elsewhere. The report highlighted that the role of the drought in destabilization is not clear, that the extent to which drought caused civil unrest is highly debated (2022, 4–53) and that studies have highlighted several other drivers (Ibid., 7–61). The IPCC described that protester demands were about politics, migrants were peripheral in the protests, and the mobilization closely resembled other Arab Spring movements. The IPCC concluded that current research "does not provide enough evidence to attribute the civil war to climate change. In contrast, it is likely that social uprisings would have occurred even without the drought" (Ibid., 16–24).

Figure 6.4 illustrates the translations and resources involved in the migration-conflict controversy. Kelley et al. based their assumptions about population growth on a UNHCR survey and data from the US Census Bureau. The presence of up to two million refugees in Syria was cited by Werrell and Femia 2012; Femia et al. 2014; Kelley et al. 2015 from the UNHCR survey, while Mills (2012) and Mohtadi (2012) did not identify sources. On water infrastructure deficiencies, both Werrell and Femia (2012) and Femia et al. (2014) referred to *Why the Water Shortages*, published on 25 March 2010 by IRIN. On other pressures, Kelley et al. quoted one farmer as saying that "the drought and unemployment were important in pushing people towards revolution" (2015, 3245). A recurring resource was a leaked US cable from the US Embassy in Damascus, which Mohtadi (2012), Gleick (2014), Friedman (2014, 2017) and Gore (2015) cited as an example of how the drought might unravel the social fabric, and how migration can "multiply" pressures and undermine stability.

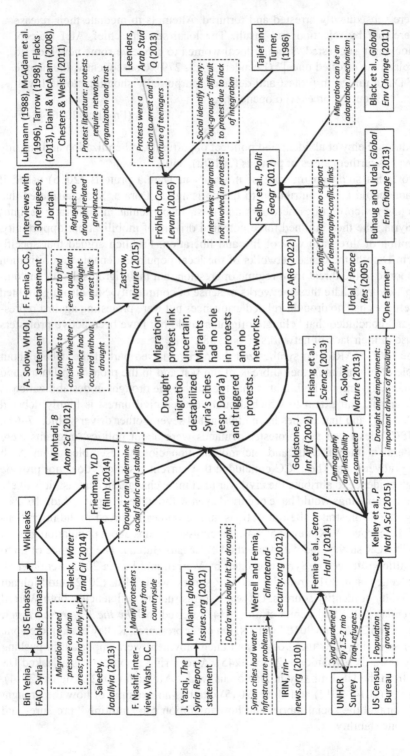

Figure 6.4 Main actors, translations and epistemic resources related to the controversy about the role of migrants in Syria's 2011 protests

To connect demography and instability, Kelley et al. referenced Goldstone (2002), Hsiang et al. (2013) and Solow (2013). Gleick's characterization of the pressures on urban areas was based on Suzanne Saleeby's article *Sowing the Seeds of Dissent* in *Jadalliya* – an electronic magazine of the Arab Studies Institute in Washington D.C. and Beirut (2014, 334, 336).

Statements about the role of rural migrants in unrests were drawn, in particular, from a CNN blog (Werrell and Femia 2012; Femia et al. 2014; Friedman 2014, 2017), Saleeby's *Jadalliya* article (Mohtadi 2012), and Friedman's interview with Nashif in *Years of Living Dangerously* (2014, 2017).

Werrell and Femia's (2012) observations about Dara'a being hard hit by the drought were based on the article *Syria: Ongoing Unrest Threatening Economy*, published on *globalissues.org* by Mona Alami, who, in turn, described the situation by quoting Jihad Yazigi, Damascus-based editor in chief of *The Syria Report*, an online business journal based in Paris. Gleick's remarks about the bread basket of Syria being crippled by drought drew on Saleeby. Finally, Friedman's observations about the role of drought refugees in Dara'a were based on his interview with Faten:

> When protests in Dara'a began in March 2011, Faten and other "drought refugees" signed on. "Since the first cry of 'Allahu akbar', we all joined . . .".
> Was this about the drought? "Of course", she said, "the drought and unemployment were important in pushing people toward revolution".
> (Friedman 2013, 2014, 2017)

De Châtel criticized Mills (2012) for inaccuracies about drought impacts in Dara'a, and drew data from the Syrian Agricultural Database to show that precipitation in Dara'a was average or above average in the winter of 08/09 and 09/10 (2014, 525). Selby's emphasis on repression and political opportunity was based on Leenders and Heydemann (2012), and Fröhlich referenced Leenders (2013) on the provocations of the local security chief.

In his criticisms, Zastrow quoted Andrew Solow, environmental statistician at the Woods Hole Oceanographic Institute in Massachusetts:

> In the absence of this drought, would there have been violence? And that's not a question you can answer . . . We don't have models like that.
> (Zastrow 2015)

Similarly, Zastrow quoted Francesco Femia, who, while emphasizing that "there was significant social unrest because of the drought", described:

> It is hard to gather even qualitative data to answer that question. Asking people if they joined the insurgency because of the drought would, in principle, provide insight, but their answers are subject to the elusiveness of collective political memory. "The movement gets mythologized and why people join changes" . . . "Memories are going to be hard to extract".
> (Ibid.)

To highlight that conflict literature does not support links between demographic change and conflict, Selby et al. (2017(a), 239) cited Urdal (2005), Buhaug and Urdal (2013) and studies that consider migration as an adaptation mechanism that helps reduce conflict risks (Black et al. 2011; Hartmann 2010). Similarly, Fröhlich referenced three publications (Reuveny 2007, 2008; Raleigh 2010; Raleigh et al. 2010), which questioned the causality between climate migration and conflict (2016, 39).

Fröhlich's study to question the role of migrants in the 2011 protests was based on 30 interviews she conducted with Syrian refugees in Azraq and Zaatari refugee camps and in Irbid and Ramtha in Jordan. She drew on the social movement theory of McAdam and Flacks to stipulate that collective action is organized via networks, and referenced Luhmann to emphasize that trust is needed for a network to lead to action because people must count on mutual support in extreme situations (2016, 44). Fröhlich considered this particularly pertinent in Syria, where, according to a refugee at the Zaatari camp, the regime said "do what you want but stay out of politics" and responded brutally to opposition (Ibid.). Drawing further on McAdam and Flacks, Fröhlich highlighted the three potential drivers of mobilizations. The interviewees told Fröhlich that in Syria, one must rely on oneself, and that in one village, the inhabitants avoided the state and the state avoided them. Based on Flack's work, Fröhlich considered that subordinated people take action when they see a threat to habitual ways, and that the "common view" was that life as it was under Hafez al-Assad came under threat during Bashar (Ibid., 45). In terms of the networks, Fröhlich described, based on Leenders, that Dara'a had dense social, economic and cultural networks, including family clan structures, which enabled mobilization (Ibid.). Following Tajlef and Turner's social-identity theory, she elaborated how the migrants were considered an "out-group", and an interviewee from Dara'a stated that migrants from Hasakah had nothing to do with politics. Other interviewees described that the migrants left when the protests began, thought they were pro-regime agents, and stated that they were 90 per cent pro-Assad because many of them or their friends were in the army (Ibid.). Based on these theoretical sources and interviews, Fröhlich concluded that opportunities, threats and aspirations toward liberation existed in Dara'a, but (climate) migrants did not start protests because they were marginalized and outside Dara'awi identity.

Selby et al. built on Fröhlich's interviews to demonstrate that drought-related grievances did not feature in the protester demands during the uprisings, and that most saw protests as reaction to repression (Selby et al. 2017(a)). Selby et al. highlighted that past contributions did not properly consider whether migrants were involved in protests, and described how interviewees stated that the migrants had no political involvement or options, and left when the protests started. They also pointed out that only one interviewee claimed that migrants were involved (Ibid.).

Finally, the IPCC review of the role of migrants in the 2011 protests covered several studies, including

- The notion that protester demands were about political issues was based on Selby et al. (2017(a)) and Ash and Obradovich (2020);

- The observation that migrants were peripheral in the 2011 protests drew on Fröhlich (2016);
- The argument that the Arab spring was a key factor was cited from Leenders (2013).

Climate change facilitated the rise of ISIS in Syria

Finally, the Contributions discussed the linkages between climate change and ISIS in particular in four ways:

- Higher temperatures and violence are correlated. Year 2014 was unusually hot in Iraq and Syria. Thus, heat and the rise of ISIS might be connected (Holthaus 2014; France 24 2014).
- Climate change played a key role in the drought that triggered the civil war. In the chaos that ensued, ISIS "stole onto the scene". Thus, climate change created the conditions for the rise of ISIS (Berkell 2014; Bremmer 2015; Kapur 2015).
- The agricultural crisis caused by the climate-induced drought drove people into hunger and desperation, which enabled ISIS recruitment (Brand 2015; Schwartzstein 2019; Varfolomeeva 2019; Granger 2019).
- Operation areas of ISIS corresponded to areas impacted by the drought (Granger 2019).

Most contributions linking climate and ISIS did not trigger controversies, but criticisms were immediate when Democratic Presidential candidate O'Malley discussed the link in an interview. Republican National Committee (RNC) Chairman Reince Priebus commented on Fox News:

> Whether it's the weak Obama-Clinton nuclear deal that paves the way for Iran to obtain an atomic bomb or Martin O'Malley's absurd claim that climate change is responsible for ISIS, it's abundantly clear no one in the Democratic Party has the foreign policy vision to keep America safe.
>
> (Kapur 2015)

Criticisms also appeared on the webpage *newsbusters.org*, where Tom Blumer observed that the "leftist" news media ignored O'Malley's "embarrassing" comments, and that

> As to farming, well, that depends on things like reliable water supplies and supply chains. It seems like a non-stop atmosphere of terror and tyranny, which has been around for almost a half-century now since the 1967 and 1973 Arab wars with Israel, would have a lot more to do with why farmers can't make a living than the rise of ISIS in the past few years . . . But maybe Martin O'Malley possesses special knowledge about all of this that the rest of us don't have, and can prove that climate change is really to blame for this. I doubt it.
>
> (Blumer 2015)

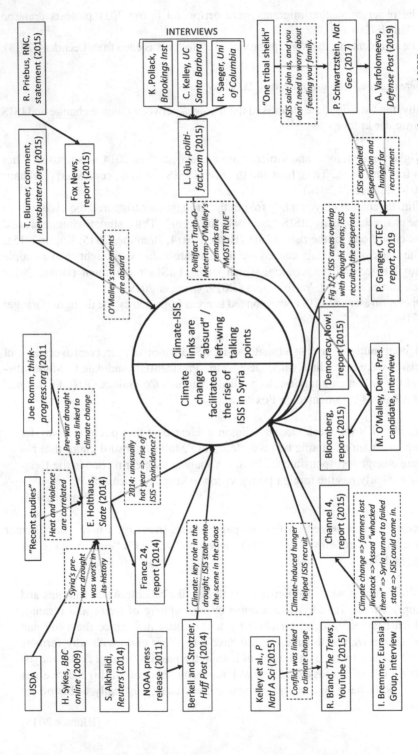

Figure 6.5 Main actors, translations and epistemic resources related to the controversy about the links between climate change and ISIS

Following this controversy, the modality of comparison of evidence was also applied by the fact-checking website *politifact.com*, operated by the *Tampa Bay Times*. It assessed O'Malley's claims in light of the criticisms (Qiu 2015), concluding that they reflected expert opinions about the role of droughts in the conflict and were not overstated. Thus, O'Malley's claim is rated "mostly true" on the Politifact Truth-O-Meter[tm] scale (used to indicate the truthfulness of statements on a scale ranging from "true" to "pants on fire") (Qiu 2015).

Figure 6.5 identifies the actors, resources and translation relevant to climate-ISIS links. The resources included NOAA's 2011 press release on the results of Hoerling et al., as well as Kelley et al.'s findings about the climate-drought link (Brand 2015; Kapur 2015; Qiu 2015). For the correlation between heat and violence, Holthaus (2014) referenced "recent studies", and the notion that climate-induced poverty provided ISIS with recruitment opportunities was drawn from Schwartzstein's article in *National Geographic*, and Varfolomeeva's article in *Defense Post*, both of which cited an interview with "one tribal sheikh" (2019). In addition, Qiu's review of the controversy between O'Malley and conservative commentators involved interviews with researchers, mainly the authors of Kelley et al. 2015, and the comments by Preibus and Blumer aimed at refuting O'Malley's statements about climate-ISIS links did not identify any visible epistemic resources.

This chapter has described the controversies, translations and knowledge resources involved in the climate-Syria debate. The next chapter will consider the similarities and differences between the Darfur and Syria cases, and begin mapping the modes of ordering of the actor-networks generated by the translations described in the preceding chapters.

Notes

1 For these stations, the authors indicated that about 20 per cent of months in 2004–2012 were missing data, and they estimated these data points by assuming that they deviate from 1961–1990 average by similar percentage as other months of the rainy seasons in question.
2 De Schutter's reports also provided conflicting numbers. The press release stated that "it is estimated that some 600,000 people have migrated out of the affected regions to urban centres, on both a seasonal and semi-permanent basis" (De Schutter 2010, 1). In contrast, the mission report stated that "[a]s a result of the repeated droughts, many families migrated to the urban centres, in the hope of finding seasonal or more permanent employment: widely cited estimates are that, in 2009, 29 to 30,000 families migrated, and that the figure in 2010 would be higher, approximating 50,000 families." (De Schutter 2011, 5).
3 According to Selby et al. (2017(a)), in May 2008, fuel prices increased by 342 per cent, and fertilizer prices by 200 to 450 per cent.
4 Selby et al. highlighted that protest opportunities were created when Assad moved security forces to Kurdish areas, and the proximity of Dara'a to Jordan provided access to networks of arms, funding and escape routes (2017(a), 240).

Literature

Adelphi 2015: *Climate Change and the Syrian Civil War (ECC Factbook Conflict Analysis)*. Video published on YouTube. 11 November 2015. At: www.youtube.com/watch?v=ZJWbxUUi4ME (accessed 5 March 2021).

Angermayer, G.; Dinc, P. and Eklund, L. 2022: *The Syrian Climate-Migration-Conflict Nexus: An Annotated Bibliography*. Center for Advanced Middle Eastern Studies, Lund. At: www.cmes.lu.se/sites/cmes.lu.se/files/2022-03/Syrian%20Climate-Migration-Conflict%20Nexus.pdf (accessed 4 August 2023).

Berkell, K.A. 2014: *How climate change helped ISIS*. Huffington Post. 29 November. At: www.huffpost.com/entry/how-climate-change-helped_b_5903170 (accessed 20 February 2022).

Blumer, T. 2015: *Press loved O'Malley's 'climate change plan' in June, ignores his linking it to the rise of ISIS now*. Newsbusters.org. 21 July (accessed 5 August 2022).

Brand, R. 2015: *ISIS vs. climate change – which kills more?* The Trews. 5 March. At: www.youtube.com/watch?v=Zrr5BvrAo-Y (accessed 15 May 2021).

Bremmer, I. 2015: *Interview at Channel 4 News*. 16 June. At: www.facebook.com/Channel 4News/videos/the-syria-war-began-because-of-climate-changeso-said-political-scientist-ian-bre/10153008539311939/ (accessed 28 July 2022).

CNA Military Advisory Board 2014: *National Security and the Accelerating Risks of Climate Change*. CNA Corporation, Alexandria, VA. Available at: www.cna.org/cna_files/pdf/MAB_5-8-14.pdf (accessed on 23 November 2019).

Daoudy, M. 2019: *The Origins of the Syrian Conflict: Climate Change and Human Security*. Cambridge University Press, Cambridge.

Daoust, G. and Selby J. 2022: *Understanding the politics of climate security policy discourse: The case of the Lake Chad Basin*. In: Geopolitics. 2 January. At: www.tandfonline.com/doi/full/10.1080/14650045.2021.2014821.

De Châtel, F. 2014: *The role of drought and climate change in the Syrian uprising: Untangling the triggers of the revolution*. In: Middle Eastern Studies, 50(4), 521–35.

De Schutter, O. 2010: *Two to three million Syrians face food insecurity, according to UN food expert*. Press release, 7 September. At: http://www.srfood.org/images/stories/pdf/press_releases/20100907_press-release_syria-mission_en.pdf (accessed 29 July 2022).

De Schutter, O. 2011: *Report of the special rapporteur on the right to food, Olivier De Schutter, to the UN General Assembly*. Human Rights Council. A/HRC/16/49/Add.2.

The Economist 2015: *Warriors and weather: Climate change and national security in America*. The Economist. 16 December. At: www.dailymotion.com/video/x6fbmah (accessed 25 November 2019).

Femia, F.; Sternberg, T. and Werrell, C. 2014: *Climate hazards, security, and the uprisings in Syria and Egypt*. In: Seton Hall Journal of Diplomacy and International Relations, XVI(1), 71–84.

Fischetti, M. 2015: *Climate change hastened Syria's civil war*. Scientific American. 2 March. At: www.scientificamerican.com/article/climate-change-hastened-the-syrian-war/ (accessed 12 November 2020).

France 24 2014: *ISIS and Climate Change*. 30 June. At: www.youtube.com/watch?v=_WFqpEBiZ04 (accessed 14 May 2021).

Friedman, T. 2012: *The Other Arab Spring*. The New York Times. 7 April. At: www.nytimes.com/2012/04/08/opinion/sunday/friedman-the-other-arab-spring.html (accessed 27 June 2022).

Friedman, T. 2013: *Without water, revolution*. The New York Times. 18 May. At: www.nytimes.com/2013/05/19/opinion/sunday/friedman-without-water-revolution.html (Accessed 23 November 2019).

Friedman, T. 2014: In: *Years of Living Dangerously, Episode 1: Dry Season*. 13 April. Showtime. At: www.amazon.com/Years-Living-Dangerously-Complete-Showtime/dp/B00NDOXDJU (accessed 23 July 2021).

Friedman, T. 2017: *Climate Wars – Syria.* 9 February. At: www.youtube.com/watch?v= i31v1z-3Z8 (accessed 28 July 2022).

Fröhlich, C. 2016: *Climate migrants as protestors? Dispelling misconceptions about global environmental change in pre-revolutionary Syria.* In: Contemporary Levant, 1(1), 38–50.

Gleick, P. 2014: *Water, drought, climate change, and conflict in Syria.* In: Weather, Climate and Society, 6(3), 331–40.

Gleick, P. 2017: *Climate, water, and conflict: Commentary on Selby et al. 2017.* In: Political Geography, 60, 248–50.

Gore, A. Jr. 2015: *Speech at the 2015 Climate Summit of the Americas.* 7 July. At: www. ontario.ca/page/climate-summit-americas-retrospective (accessed 19 July 2021).

Granger, P. 2019: *A Perfect Storm: How Climate Change Contributed to the Rise of the Islamic State.* Center on Terrorism, Extremism, and Counterterrorism. Middlebury College, Monterey, CA. At: www.middlebury.edu/institute/sites/www.middlebury.edu.institute/files/2019-08/Paula%20Granger%20CTEC%20Paper.pdf?fv=-ym_XGjc (accessed 5 August 2022).

Gray, R. 2015: *Did climate change trigger the war in Syria? Severe drought may have contributed to uprising, study reveals.* Daily Mail. 2 March. At: www.dailymail.co.uk/sciencetech/article-2976063/Did-climate-changetrigger-war-Syria-Severe-drought-contributed-uprising-study-reveals.html (accessed 23 November 2019).

Hendrix, C.S. 2017: *A comment on 'climate change and the Syrian civil war revisited'.* In: Political Geography, 60(1), 251–2.

Hoerling, M.; Eischeid, J.; Perlwitz, J.; Quan, X.; Zhang, T. and Pegion, P. 2012: *On the increased frequency of Mediterranean drought.* In: Journal of Climate. March. At: https://journals.ametsoc.org/doi/full/10.1175/JCLI-D-11-00296.1 (accessed 27 August 2019).

Holthaus, E. 2014: *Is climate change destabilizing Iraq?* Slate. 27 June. At: https://slate.com/technology/2014/06/isis-water-scarcity-is-climate-change-destabilizing-iraq.html (accessed 14 May 2021).

Ide, T. 2018: *Climate war in the Middle East? Drought, the Syrian civil war and the state of climate-conflict research.* In: Current Climate Change Reports, 4(4), 347–54.

IPCC 2022: (Pörtner, H.-O.; Roberts, D.C.; Tignor, M.; Poloczanska, E.S.; Mintenbeck, K.; Alegría, A.; Craig, M.; Langsdorf, S.; Löschke, S.; Möller, V.; Okem, A. and Rama, B. (Eds.)): *Climate Change 2022: Impacts, Adaptation, and Vulnerability. Contribution of Working Group II to the Sixth Assessment Report of the IPCC.* Cambridge University Press. In Press. At: www.ipcc.ch/report/sixth-assessment-report-working-group-ii/ (accessed 19 June 2022).

IRIN 2009: *Syria: Drought Response Faces Funding Shortfall.* 24 November. At: https://reliefweb.int/report/syrian-arab-republic/syria-drought-response-faces-funding-shortfall (accessed 29 July 2022).

Kapur, S. 2015: *Martin O'Malley Clashes With GOP Over Link Between Climate Change, Islamic State.* Bloomberg. 22 July. At: www.bloomberg.com/news/articles/2015-07-22/o-malley-clashes-with-republicans-over-link-between-climate-change-and-isis (accessed 5 August 2022).

Kelley, C.P.; Mohtadi, S.; Cane, M.A.; Seager, R. and Kushnir, Y. 2015: *Climate change in the fertile crescent and implications of the recent Syrian drought.* In: PNAS, 112(11), 3241–6.

Kelley, C.P.; Mohtadi, S.; Cane, M.A.; Seager, R. and Kushnir, Y. 2017: *Commentary on the Syria case: Climate as a contributing factor.* In: Political Geography, 60(1), 245–7.

Kerry, J. 2015: *Remarks by the Secretary of State at Old Dominion University on Climate Change and National Security.* 10 November. At: https://2009-2017.state.gov/secretary/remarks/2015/11/249393.htm (accessed 25 November 2019).

Mills, R. 2012: *Long drought that helped to spark an uprising in Syria*. The National. 27 July. At: www.thenationalnews.com/business/long-drought-that-helped-to-spark-an-uprising-in-syria-1.387746 (accessed 27 July 2022).

Mohtadi, S. 2012: *Climate change and the Syrian uprising*. In: Bulletin of the Atomic Scientists. 16 August. At: https://thebulletin.org/2012/08/climate-change-and-the-syrian-uprising/#post-heading (accessed 30 April 2023).

NPR 2013: *How Could a Drought Spark a Civil War?* 8 September. At: www.npr.org/2013/09/08/220438728/how-could-a-drought-spark-a-civil-war?t=1658950390329 (accessed 27 July 2022).

Obama, B. 2015: *Remarks by the President at the United States Coast Guard Academy Commencement*. 20 May. At: https://obamawhitehouse.archives.gov/the-press-office/2015/05/20/remarks-president-united-states-coast-guard-academy-commencement (accessed 25 November 2019).

Phillips, C. 2020: *The Battle for Syria: International Rivalry in the Middle-East*. Yale University Press, New Haven.

Qiu, L. 2015: *Fact-checking the link between climate change and ISIS*. Politifact.com. 23 September. At: www.politifact.com/factchecks/2015/sep/23/martin-omalley/fact-checking-link-between-climate-change-and-isis/ (accessed 5 August 2022).

Schwartzstein, P. 2019: *Climate Change and Water Woes Drove ISIS Recruiting in Iraq*. At: www.nationalgeographic.com/science/article/climate-change-drought-drove-isis-terrorist-recruiting-iraq (accessed 5 August 2022).

Selby, J.; Dahi, O.; Fröhlich, C. and Hulme, M. 2017(a): *Climate change and the Syrian civil war revisited*. In: Political Geography, 60(1), 232–44.

Selby, J.; Dahi, O.; Fröhlich, C. and Hulme, M. 2017(b): *Climate change and the Syrian civil war revisited: A rejoinder*. In: Political Geography, 60(1), 253–5.

UNHCR 2011: *Report of the Special Rapporteur on the Right to Food on His Mission to Syria, Addendum*. A/HRC/16/49/Add.2. 27 January. At: https://www2.ohchr.org/english/bodies/hrcouncil/docs/16session/A.HRC.16.49.Add.2_en.pdf (accessed 28 July 2022).

UNOCHA 2010: *Syria drought response plan 2009–2010 mid-term review*. New York. At: https://reliefweb.int/attachments/1733ced5-9579-3bc6-a4d3-3c53ba43793a/20E00ADAF9F3C153852576D20068E86B-Full_Report.pdf (accessed 29 July 2022).

Wendle, J. 2016: *Syria's climate refugees*. In: Scientific American, 314(3), 50–5.

Werrell, C.E. and Femia, F. 2012: *Syria: Climate Change, Drought and Social Unrest*. CCS. 29 February 2012. At: https://climateandsecurity.org/2012/02/syria-climate-change-drought-and-social-unrest/ (accessed 10 November 2020).

Werrell, C.E. and Femia, F. 2017: *The Climate Factor in Syrian Instability: A Conversation Worth Continuing*. CCS. 8 September. at: https://climateandsecurity.org/2017/09/08/the-climate-factor-in-syrian-instability-a-conversation-worth-continuing/ (accessed 28 July 2022).

Werrell, C.E.; Femia, F. and Sternberg, T. 2015: *Did we see it coming?: State fragility, climate vulnerability, and the uprisings in Syria and Egypt*. In: SAIS Review of International Affairs, 35(1), 29–46.

YCC 2015: *Drought, Water, War, and Climate Change*. At: www.youtube.com/watch?v=BbkNcvGHZwc (accessed 15 May 2023).

Yamin, F. 2019: *This is the only way to tackle the climate emergency*. Time Magazine, 14 June. At: https://time.com/5607152/extinction-rebellion-farhana-yamin/ (accessed 11 November 2020).

Zastrow, M. 2015: *Climate change implicated in current Syrian conflict*. In: Nature. 2 March. At: www.nature.com/news/climate-change-implicated-in-current-syrian-conflict-1.17027 (accessed 27 April 2021).

Additional literature referenced by the contributions to the climate-Syria debate

Ababsa, M. 2015: *The end of a world: Drought and agrarian transformation in North-east Syria (2007–2010)*. In: Hinnebusch, R. and Zintl, T. (Eds.): *Political Economy and International Relations: Vol. 1. Syria from Reform to Revolt*. Syracuse University Press, Syracuse, 199–222.

Aïta, S. 2009: *Labour Markets Performance and Migration Flows in Syria: National Background Paper, Labour Markets. Performance and Migration Flows in Arab Mediterranean Countries: Determinants and Effects*. At: http://citeseerx.ist.psu.edu/viewdoc/download?doi=10.1.1.472.2188&rep=rep1&type=pdf (accessed 2 August 2022).

Al-Hindi, A. 2011: *Syria's agricultural sector: Situation, role, challenges and prospects*. In: Hinnebusch, R. et al. (Eds.): *Agriculture and Reform in Syria*. University of St Andrew Centre for Syrian Studies, St Andrew, 15–55.

Ash, K.; and Obradovich, N.: 2020: *Climatic stress, internal migration, and Syrian civil war onset*. In: Journal of Conflict Resolution, 64(1), 3–31.

Becker, A.; Finger, P.; Meyer-Christoffer, A.; Rudolf, B.; Schamm, K.; Schneider, U. and Ziese, M. 2013: *A description of the global land-surface precipitation data products of the global precipitation climatology centre with sample applications including centennial (trend) analysis from 1901 present*. In: Earth Systems Science Data, 5(1), 71–99.

Bellprat, O. and Doblas-Reyes, F. 2016: *Attribution of extreme weather and climate events overestimated by unreliable climate simulations*. In: Geophysical Research Letters, 43(5), 2158–64.

Bergaoui, K. 2015: *The contribution of human-induced climate change to the drought of 2014 in the southern Levant region*. In: Bulletin of the American Meteorological Society, 96(12), 66.

Black, R.; Bennett, S.R.G.; Thomas, S.M. and Beddington, J.R. 2011: *Climate change: migration as adaptation*. In: Nature, 478, 447–9.

Breisinger, C.; Ecker, O.; Al-Riffai, P. and Yu, B. 2012: *Beyond the Arab Awakening: Policies and Investments for Poverty Reeducation and Food Security*. IFPRI, Washington, DC.

Buhaug, H. and Urdal, H. 2013: *An urbanization bomb? Population growth and social disorder in cities*. In: Global Environmental Change, 23(1), 1–10.

Chalcraft, J. 2008: *The Invisible Cage: Syrian Migrant Workers in Lebanon*. Stanford University Press, Stanford.

Chesters, G. and Welsh, I. 2011: *Social Movements: The Key Concepts*. Routledge, London.

Cook, B.I.; Anchukaitis, K.J.; Touchan, R.; Meko, D.M. and Cook, E.R. 2016: *Spatiotemporal drought variability in the Mediterranean over the last 900 years*. In: Journal of Geophysical Research: Atmospheres, 121(5), 2060–74.

Cook, E.; Seager, R.; Kushnir, Y.; Briffa, K.; Büntgen, U.; Frank, D. et al. 2015: *Old world megadroughts and pluvials during the common era*. In: Science Advances, 1(10), 1500561.

Diani, M. and McAdam, D. (Eds.) 2003: *Social Movements and Networks: Relational Approaches to Collective Action*. Oxford University Press, Oxford.

Dong, B. and Sutton, R. 2015: *Dominant role of greenhouse-gas forcing in the recovery of Sahel rainfall*. In: Nature Climate Change, 5(8), 757–60.

Eklund, L. and Thompson, D. 2017: *Differences in resources management affects drought vulnerability across the borders between Iraq, Syria, and Turkey*. In: Ecology an Society, 22.

Goldstone, J. 2002: *Population and security: How demographic change can lead to violent conflict*. In: Journal of International Affairs, 56(1), 3–22.

Hartmann, B. 2010: *Rethinking climate refugees and climate conflict: Rhetoric, reality and the politics of policy discourse*. In: Journal of International Development, 22(2), 233–46.

He, J. and Soden, B. 2017: *A re-examination of the projected subtropical precipitation decline*. In: Nature Climate Change, 7(1), 53–7.

Hsiang, S.M.; Burke, M.B. and Miguel, E. 2013: *Quantifying the influence of climate on human conflict*. In: Science, 341, 1235367.

Hulme, M. 1996: *Recent climate change in the world's drylands*. In: Geophysical Research Letters, 23(1), 61–4.

Hulme, M. 2001: *Climatic perspectives on Sahelian desiccation: 1973–1998*. In: Global Environmental Change, 11, 19–29.

Hulme, M. 2014: *Attributing weather extremes to "climate change": A review*. In: Progress in Physical Geography, 38(4), 499–511.

IPCC 2018 (Masson-Delmotte, V.; Zhai, P.; Pörtner, H.-O.; Roberts, D.; Skea, J.; Shukla, P.R.; Pirani, A.; Moufouma-Okia, W.; Péan, C. Pidcock, R.; Connors, S.; Matthews, J.B.R.; Chen, Y.; Zhou, X.; Gomis, M.I.; Lonnoy, E.; Maycock, T.; Tignor, M. and Waterfield, T. (Eds.): Global Warming of 1.5°C. An IPCC Special Report on the Impacts of Global Warming of 1.5°C Above Pre-Industrial Levels and Related Global Greenhouse Gas Emission Pathways, in the Context of Strengthening the Global Response to the Threat of Climate Change, Sustainable Development, and Efforts to Eradicate Poverty. Cambridge University Press, Cambridge and New York. At: www.ipcc.ch/sr15/ (accessed 5 July 2021).

Kelley, C.; Ting, M.; Seager, R. and Kushnir, Y. 2012: *Mediterranean precipitation climatology, seasonal cycle, and trend as simulated by CMIP5*. In: Geophysical Research Letters, 39(29), 21703.

Kitoh, A.; Yatagai, A. and Alpert, P. 2008: *First super-high-resolution model projection that the ancient 'fertile crescent' will disappear in this century*. In: Hydrological Research Letters, 2, 1–4.

Leenders, R. 2012: *Collective action and mobilization in Dar'a: An anatomy of the onset of Syria's popular uprising*. In: Mobilization, 17(4), 419–34.

Leenders, R. 2013: *Social movement theory and the onset of the popular uprising in Syria*. In: Arab Studies Quarterly, 35(3), 273–89.

Leenders, R. and Heydemann, S. 2012: *Population mobilization in Syria: Opportunity and threat, and the social networks of the early risers*. In: Mediterranean Politics, 17(2), 139–59.

Luhmann, N., 1988: *Familiarity, confidence, trust: Problems and alternatives*. In: Gambetta, D. (Ed.): Trust: Making and Breaking Cooperative Relations. Basil Blackwell, Oxford.

Massoud, A. 2010: *Years of Drought: A Report on the Effects of Drought on the Syrian Peninsula*. Heinrich Böll-Stiftung. Berlin. At: https://lb.boell.org/en/2010/12/14/years-drought-report-effects-drought-syrian-peninsula (accessed 22 December 2018).

McAdam, D.; McCarthy, J.D., and Zald, M.N. 1996: *Comparative Perspectives on Social Movements: Political Opportunities, Mobilizing Structures, and Cultural Framings*. Cambridge Studies in Comparative Politics. Cambridge University Press, Cambridge.

Nabhan, G. 2010: *Drought drives Middle Eastern pepper farmers out of business, threatens prized heirloom chillies*. Grist.org. At: http://grist.org/food/2010-01-15-drought-drives-middle-eastern-peppers (accessed 5 August 2022).

Raleigh, C. 2010: *Political marginalization, climate change and conflict in Africa Sahel states*. In: International Studies Review, 12, 69–86.

Raleigh, C.; Linke, A.; Hegre, H. and Karlsen, J. 2010: *Introducing ACLED: An armed conflict location and event dataset*. In: Journal of Peace Research, 47, 651–60.

Reuveny, R. 2007: *Climate change-induced migration and violent conflict*. In: Political Geography, 26, 656–73.

Reuveny, R. 2008: *Ecomigration and violent conflict: Case studies and public policy implications*. In: Human Ecology, 36, 1–13.

Saleeby, S. 2012: *Sowing the seeds of dissent: Economic grievances and the Syrian social contract's unravelling*. Jadaliyya. 16 February. At: www.jadaliyya.com/pages/index/4383/sowing-the-seeds-of-dissent_economic-grievances-an (accessed 15 July 2022).

Selby, J. 2019: *Climate change and the Syrian civil war, Part II: The Jazira's agrarian crisis*. In: Geoforum, 101, 260–74.

Solow, A.R. 2013: *Global warming: A call for peace on climate and conflict*. In: Nature, 97(7448), 179–80.

Sousa, P.N.; Trigo, R.M.; Aizpurua, P.; Nieto, R.; Gimeno, L. and Garcia-Herrera, R. 2011: *Trends and extremes of drought indices throughout the 20th century in the Mediterranean*. In: Natural Hazards Earth Systems Science, 11, 33–51.

Tarrow, S. 1998: *Power in Movement: Social Movements and Contentious Politics*. Cambridge University Press, Cambridge.

Tertrais, B. 2011: *The climate wars myth*. In: The Washington Quarterly, 34(3), 17–29.

Trigo, R.; Gouveia, C.M. and Barriopedro, D. 2010: *The intense 2007–2009 drought in the fertile crescent: Impacts and associated atmospheric circulation*. In: Agricultural and Forest Meteorology, 150(9), 1245–57.

Urdal, H. 2005: *People versus Malthus: Population pressure, environmental degradation, and armed conflict revisited*. In: Journal of Peace Research, 42(4), 417–34.

Varfolomeeva, A. 2019: *Climate change: An increasingly dangerous recruiting sergeant for insurgent groups*. The Defense Post. 13 April. At: www.thedefensepost.com/2019/04/13/climate-change-insurgency/ (accessed 5 August 2022).

Worth, R. 2010: *Earth is parched where Syrian farms thrived*. The New York Times. 13 October. At: www.nytimes.com/2010/10/14/world/middleeast/14syria.html (accessed 15 May 2023).

Zappa, G.; Hawcroft, M.K.; Shaffrey, L.; Black, E. and Brayshaw, D.J. 2015: *Extratropical cyclones and the projected decline of Winter Mediterranean precipitation in the CMIP5 models*. In: Climate Dynamics, 45, 1727–38.

Zappa, G., Hoskins, B.J. and Shepherd, T.G. 2015: *The dependence of wintertime Mediterranean precipitation on the atmospheric circulation response to climate change*. In: Environmental Research Letters, 10, 104012.

7 Comparing the two climate wars

Mapping actors, translations and knowledge resources

Chapters 5 and 6 described the empirical results of "following the actors" – the first analytical principle of the research strategy of this book outlined in Chapter 4 – in order to map their efforts to connect or disconnect climate change and the conflicts in Darfur and Syria. The chapters briefly described both conflicts and provided a chronological overview of the climate-conflict debates. Following the second analytical principle – focusing on controversies – the mapping of translations began by first identifying key controversies within both climate-conflict debates. These controversies provided a helpful entry point to the debate because they incentivized actors to intensify translations and efforts to mobilize knowledge resources (Callon 1986).

The key controversies were as follows:

- For Darfur, actors debated whether (a) Darfur was the "first climate war"; (b) Darfur had climate-induced rain failure prior to the conflict; (b) lack of water and land was a conflict driver; (d) farmers and pastoralists began to fight; (e) Darfur provides a warning about global resource wars, and; (f) climate-Darfur links reduce Khartoum's responsibility.
- For Syria, the controversies were about whether: (a) the 2006–2009 drought was the worst in Syrian history; (b) climate change and the drought were connected; (c) the drought played a major role in rural-urban migration; (d) migration destabilized Syrian cities and led to war, and (e) climate change facilitated the rise of ISIS.

These controversies provided the structure for the mapping of translations. The identification of controversies was followed by descriptions of how actors articulated initial translations to advance climate-conflict links, how others outlined dissenting translations in response, and what knowledge resources were mobilized to support both sides of the controversies.

In accordance with the further analytical principles identified in Chapter 4, the descriptions in Chapter 5 and 6 sought to avoid (a) any conscious efforts to settle the controversies, but rather to reflect the expressions actors give to climate-conflict

DOI: 10.4324/9781003451525-7

links; (b) any determination of which actors have a legitimate voice in the debate; (c) speculating about social forces or motivations of actors to explain the climate-conflict debates. In addition, as emphasized by the fifth analytical principle – the accounts in Chapters 5 and 6 are only a partial "snapshot" of the visible aspects of the climate-conflict debates and aimed at providing descriptions that reflect the activities of actors, the relations between them, and the organization of their associations. Most likely, as emphasized by Latour (2005, 201) a great amount of relevant translations and knowledge resources were operating in the background of the debate, but could not be mapped due to lack of information, space and time.

Based on the mapping, this chapter will begin to describe the basic structure of the actor-networks emerging from the translations and mobilization of knowledge resources. As discussed in Chapter 4, the results of translations are new associations between actors, which constitute actor-networks – enabling and constraining arrangements of actors. The modes of ordering of these actor-networks – meaning the constraining and enabling influences of the associations that actors have put in place when debating the links between climate change and the two conflicts – will be the subject of the next four chapters. This chapter will focus, in particular, on three fundamental characteristics of the actor-networks:

- How the core of both climate-conflict actor-networks is formed by articulations of linear causal mechanisms between climate and the two conflicts;
- How both climate-conflict debates progress through three overlapping stages characterized by changing modalities of translation: initial articulations of the linear causal mechanisms, science-based reviews of the mechanism components, and efforts to review and synthesize existing evidence;
- The diversity of actors, knowledge resources and the channels of circulation of translations involved in the debates.

Linear causal mechanisms: connecting and structuring

The most intensively debated parts of the controversies are the components of linear causal mechanisms that actors established to connect climate change and the two conflicts. The climate-Darfur mechanism encompasses the following components: (1) climate change caused rainfall decline in the region; (2) lack of rainfall led to resource scarcity; and (3) resource scarcity caused tensions between farmers and pastoralists, and those tensions escalated into the conflict in 2003. The mechanism connecting climate and Syria has four main components: (1) Prior to conflict, Syria had its worst drought ever; (2) the drought was made more likely by climate change; and (3) the drought caused a massive internal migration in Syria; 4. The migration destabilized Syrian cities and lead to protests and war. Other studies (Ide 2018; Selby et al. 2022) have observed similar ways of making the connections, and noted how the connections between climate and conflict tend to be built to be "intuitively sensible" (Verhoeven 2011, 680).

These mechanisms are a consequence of the translations by actors. They form the core of the climate-conflict actor-networks by providing logically coherent causal connections between climate and the conflicts through a sequence of causal steps between climate impacts and conflict outbreaks. In other words, they explain why climate change and the conflicts are linked. The mechanisms also have a second function: they evolve into the structure of the climate-conflict debates. The mechanisms are articulated at early stages of the debates by a small number of actors. As the debates advance, other actors structure their contributions around the components of the mechanisms, thus following the established structure.

Evolving modalities of translation

And what are the modalities through which actors try to settle the above controversies and, in particular, debate the components of the linear causal mechanisms? For both conflicts, the modalities of translation evolved through three stages: (1) initial narratives of climate-conflict links; (2) science-based criticisms or consolidations of those links; (3) comparison and review of evidence.

For both Darfur and Syria, the climate-conflict debates begin when actors introduce narratives of the linear causal mechanisms outlined above, and other aspects of climate-conflict links. The early articulations are characterized by logically coherent causal narratives accompanied by information in particular from interview statements, anecdotal evidence, and references to related scientific studies. For Darfur, the mechanism was outlined from 2004 onward. The first articulations of the climate-Syria mechanism were published in 2012. The mechanisms articulated in this stage, while both standing on a thin knowledge basis, provided the structure for the next stage.

In the second stage, the narratives of the mechanisms were followed in particular by technical questioning or supporting of the mechanism components, consolidating the role of the mechanisms as the structure of the debate. The types of knowledge resources in this stage were in strong contrast with the interview-based and anecdotal resources of the first stage. Technical elements were increasingly mobilized as the controversies intensified and actors sought resources to support their case. On Darfur, peer-reviewed scientific studies were exclusively critical of climate-conflict links, questioning whether rains really failed, whether there was resource scarcity in Darfur prior to the conflict, whether farmer-pastoralist relations are foundational elements of Darfurian society, and whether climate conflicts are really happening across Africa.

In the case of Syria, a similar technical scrutiny took place, but in this case, peer-reviewed studies both supported and criticized the components of the mechanism. Studies described drought severity, drought-climate links, the role of water and climate in Syria's deterioration, and role of migrants in the 2011 protests. Several studies focused on the individual components, while for example the study by Kelley et al. (2015) systematically mobilized evidence to support all components of the climate-Syria mechanism. Others questioned the climate-Syria mechanism by highlighting the importance of other conflict drivers than the drought, by challenging the role of migrants as drivers of protests, and by reconsidering issues of drought severity and migration levels in pre-war Syria. This second stage was characterized by the ways in which technical resources were mobilized to criticize or support the components of the mechanism articulated in the first stage. In addition, the climate-Syria debate involved an additional stage of consolidation, where in particular Kelley et al.'s findings about drought severity, role of climate change, migration and subsequent unrest were cited on several media platforms and by policymakers, including US President Obama. This was a further contrast to Darfur, where the climate-conflict debate did not recover its initial momentum after the critical contributions.

The third stage was characterized by efforts to review, compare and synthesize existing evidence and draw conclusions about climate-conflict links. While several individual researchers (Mazo 2010; Ide 2018; Angermayer et al. 2022) provided summaries of the debate, the most systematic efforts were undertaken by the IPCC through its formal syntheses of evidence for climate-Darfur and climate-Syria links in 2014 and 2022, respectively. In both cases, the IPCC concluded that there is insufficient evidence to link climate and the conflicts, that other factors were probably more important, and that the conflicts would have been likely even without climate impacts. This last stage thus involved two modalities of synthesis: efforts by individual researchers, as well as a formal institutionalized review by the IPCC.

Diverse actors, knowledge resources and circulation modalities

Throughout these three stages, a wide range of actors, knowledge resources and circulation modalities were involved in the debates.

Initial narratives

The initial narratives and criticisms of climate-Darfur links were articulated in particular by government representatives (especially the UK and Sudan), high-level UN officials, including the secretary-general, journalists, researchers, as well as international and non-governmental organizations. The statements circulated in particular in British and American newspapers. They were reflected in speeches at diverse places such as Chatham House, the Nobel ceremony and Lehigh University in Pennsylvania. They were also captured in reports by UNEP and the NGO Tearfund, and De Waal and Homer-Dixon debated causality on the website *African Arguments*. At a later stage, climate-Darfur narratives were debated in a 2020 US presidential debate, by the right-wing media outlet Breitbart News and by unexpected actors such as the GWR.

Some actors outlined their narratives by simply labeling Darfur as a "climate war" without specific resources. Others cited anonymous sources, and yet others mobilized specific resources. For example:

- The notion that rains failed before the war was built on data distributed by SMA from three rainfall stations in Darfur, on a 2013 study led by Alessandra

Giannini on links between Indian Ocean temperature and Sahelian rainfall, and on unspecified "UN statistics".
- The resource scarcity aspect was built in particular on statements by practitioners, references to the 2003 Pentagon report by Schwartz and Randall and a workshop in Khartoum.
- The farmer-pastoralist link was drawn in particular from a story told by Alex De Waal to Stephan Faris about his meeting with Sheikh Abdalla in Darfur in 1985.
- The implications of climate-conflict links beyond Darfur were considered by citing Miguel et al.'s 2004 study on connections between resource scarcity and conflicts in Africa, and UNEP and Tearfund cited climate models operated by KNMI and ILRI to predict crop losses across Africa.
- The notion that the climate-Darfur thesis lets Khartoum "off the hook" was based on a speech by Sudanese Ambassador Mohamad.
- The debate on the nature of causality relevant to climate-Darfur links did not involve visible resources, except a reference to unspecific "studies of complex systems".

While several contributions drew on the same knowledge resources, cross-references among them were almost non-existent. The initial articulations of climate-Darfur links were disconnected, not mutually supportive and based on limited resources. As a result, they remained fragmented, which prevented their consolidation into "black boxes" and made them easier targets for criticisms in the next stage.

The initial narratives of climate-Syria mechanisms were published in particular on the website of the CCS, as well as in media publications including the *NYT*. The following resources were particularly visible in the initial narratives:

- Early statements about climate-drought links were based on the study of Mediterranean drought by Hoerling et al. 2012, as well as on information about the study in a NOAA press release and Joe Romm's environmental blog *thinkprogress.org*.
- Information on drought severity on the ground was drawn from interviews with Syrian refugees and a quote from "one expert" in an article on the plight of Syria's pepper farmers in on *grist.org*.
- The impacts of drought were described on the basis of a report by the UN Office for Disaster Risk Reduction and UN Special Representative De Schutter's 2010 survey of agricultural impacts of drought.
- Estimates of post-drought migration levels in Syria were based mainly on UNO-CHA reports and the statement about the Syrian pre-war drought having driven one and a half million people from their homes, which originated in an interview with M. Alloush from the SPC published by IRIN.
- The impacts of migration in Syrian cities were elaborated on the basis of a UNHCR survey on Iraqi refugees, a leaked US embassy cable, an article in *Jadalliya*, and an interview with a Syrian refugee.
- Statements about how ISIS recruitment benefited from drought-induced desperation were based on an interview with one tribal sheikh in northern Iraq.

The climate-Syria debate thus begun like Darfur: actors articulated the mechanism and other aspects based on heterogenous knowledge resources. However, the two debates differed in terms of the actors, and the types of representations. While the first stage of the climate-Darfur debate involved a concerted effort by the UK government and UN actors, the Syria debate begun with contributions in particular by nongovernmental and media actors. In particular, no government or UN actors were involved in the initial contributions on Syria.

Scientific consideration of linear causal mechanisms

The studies on climate-Darfur links were published by researchers in the US and UK and appeared exclusively in three peer-reviewed journals: *Environmental Research Letters, International Journal of Remote Sensing* and *Global Environmental Change*. They were exclusively critical of climate-Darfur links. The resources mobilized by the studies included ground- and satellite-based measurement devices for monitoring rainfall and vegetation coverage, datasets that enabled organizing and circulating information captured by those devices, datasets on conflicts, the NDVI index to standardize and compare vegetation levels, climate and crop models to make projections about rainfall and agricultural productivity in the region, and statistical methods to create representations based on the data. For example, Kevane and Gray questioned rainfall deficits with the help of data from three stations in Darfur, two gridded datasets, statistical methodologies and comparisons with other evidence. Data was provided by research institutes, in particular the UEA CRU in Norwich, UK. Kevane and Gray's criticism of Africa-wide trends of climate conflicts were based on the UEA CRU datasets, PRIO/UCDP conflict data and multiple regression analysis. Brown's observations involved NDVI formula, 200 red and infrared photographs taken by AVHRR cameras on board of NOAA satellites, the Savitzky-Golay filter (a statistical tool), and comparisons with other vegetation indices. These are examples of the ways in which information about Darfur's climate past and future was captured in quantified data points by several technical entities, and could be inscribed into graphical and textual presentations on the basis of statistical methods and indices and circulated in academic journals in order to challenge the initial narratives of linear causal mechanisms.

The studies also involved a less technical and non-statistical modality, that of citing other studies. For example, Selby and Hoffmann cited books by De Waal as well as Young and Osman to argue that farmer-pastoralist dynamics are an outdated expression of the economic dynamics of Darfur. To highlight the diverse, modern and globalized character of Sudanese livelihoods, they referenced a factsheet by the organization Global Humanitarian Assistance. And they used an ICG report and an article in *African Affairs* to emphasize that government expenditure increased in Darfur prior to the conflict, indicating that underdevelopment was not an issue.

The Syrian debate involved a much higher number of technical resources (20 peer-reviewed studies in contrast to five on Darfur) and contributions that both supported and criticized climate-Syria links. The supportive studies were published in particular in the journals *Climate, Water and Society, Seton Hall Journal of*

International Relations and *PNAS*. A debate on climate-Syria links was conducted in particular in multiple articles in a 2017 issue of *Political Geography*.

The supportive studies focused in particular on drought severity, climate-drought links, high levels of migration, and the impacts of migration on Syria's cities. The following resources were particularly central:

- Considerations of drought severity involved data from gridded precipitation datasets, weather stations in Qamishli and Deir Ez-Zor, a GHCN database encompassing 25 rainfall stations, and NASA GRACE Tellus satellite. This also involved the use of regression analysis and drought indices (scPDSI and LWE).
- The climate-drought link was elaborated, in particular on the basis of Hoerling et al.'s study of Mediterranean drought, a study by Mathbout and Skaf involving the SPI and EDI drought indices and 16 CMIP5 climate models.
- Statements about migration were further consolidated citing unspecified UN estimates, IRIN's interview with M. Alloush, as well as a 2010 assessment by ICARDA on displacement, cited from Ababsa's 2015 *End of the World*.
- The impact of migrants on Syrian cities was considered on the basis of information on Syrian population growth from US Census Bureau, IRIN's 2010 report on water shortages, as well as literature on links between demography and stability.

The scientific findings also circulated beyond the original publications. In particular the study by Kelley received significant media attention and its results were cited in the *Daily Mail*, *Scientific American* and *Nature* and even in a video published by Russell Brand on his show *The Trews*. They were also discussed in speeches by President Obama and former Vice-President Al Gore.

The criticisms of the climate-Syria links were published in *Middle Eastern Studies*, *Contemporary Levant* and *Political Geography*. They involved in particular these knowledge resources:

- The severity of Syria's pre-war drought was questioned on the basis of an IFPRI paper that highlighted how normal droughts are in Syria, a UNHCR report that described rainfall recovery in 2009–2010, and new rainfall data from MAAR, GPCC and UEA CRU.
- Climate-drought links were scrutinized by using rainfall data, linear regression, and past publications that highlighted challenges with linear rainfall modeling.
- Criticisms of migration numbers (especially the one and a half million figure) were based on UNOCHA Syrian drought response plan, which referred to 300,000 people and 65,000 families and an article that highlighted multiple factors of migration.
- The role of migrants in protests was reconsidered on the basis of interviews with Syrian refugees, literature on protests and social movements, a study that described the protests as a reaction to the arrest and torture of teenagers in Dara'a, and studies that found no correlations between demographic change and conflict.

While reactions to criticisms of climate-Darfur links were limited, the supporters of climate-Syria links responded to the criticisms in *Political Geography* and on *climateandsecurity.org*. The reactions involved in particular the following points and resources:

- On climate-drought links, the respondents highlighted further studies of climate-drought links, climate model simulations, and tree ring studies which they claimed had been ignored by critics.
- On migration levels, the respondents argued that critics used migration numbers selectively and did not consider the high migration levels 6–18 months before uprisings. Instead, the respondents cited in particular the estimate of 300,000 families from Ababsa (2015) (taken from an IISD and UNOCHA reports).

This illustrates how the number of technical actors and complexity increased significantly during the second stage, with anecdotal and interview evidence replaced by measurement devices, databases, statistical methods, indices and figures. There are also regular spill-overs between scientific findings and media contributions. As a consequence of the increasing technicality, the debate became increasingly challenging to understand. This created a demand for a synthesis of evidence to simplify the debate and to provide manageable information for policymakers. This is what the next stage was about.

Comparison and review of evidence

The first attempt to review climate-Darfur links was the book *Climate Conflict* (2010), in which Mazo, in response to criticisms of the rainfall components of the climate-Darfur mechanism by Kevane and Gray, cited a 2009 study by Burke et al. that had found a link between temperature increase and conflict across Africa. This way, the modality of comparison of evidence was used to attempt to replace rainfall with temperature within the linear causal mechanism.

The IPCC's 2014 review of climate-Darfur links concluded that no sufficient evidence exists to call Darfur a climate war, and that non-climate factors were more relevant. It was based on the internationally agreed rules for IPCC assessments, past publications (Mazo's book and five peer-reviewed studies), and comments from experts and government representatives. As inputs, the IPCC's procedures prioritize peer-reviewed articles and books, as well as reports by governments and organizations, while media contributions and speeches are specifically excluded. This meant that most of the contributions to the climate-Darfur debate, which, as outlined earlier, were published on media platforms as well as in speeches and organization reports, were excluded from the 2014 review.

Efforts to review climate-Syria links were made by Ide (2018), Angermayer et al. (2022) and the IPCC (2022). Ide's review articulated four stages of the climate-Syria link: (1) A climate-induced drought in 2006–2009; (2) drought impacts in Syria's northeast; (3) migration to urban areas; (4) migration aggravated problems and led to an escalation. He described several studies, including De Châtel (2014), Gleick

(2014), Kelley et al. (2015), Werrell et al. (2015), Fröhlich (2016), Eklund and Thompson (2017) and Selby et al. (2017(a)), on these stages. Angermayer's et al. considered much of the same literature, providing a brief summary of each study, and highlighting drought-migration links as a major research gap due to lack of data, undifferentiated consideration of migration, and low attention to migration literature. The IPCC's 2022 review considered 20 peer-reviewed studies, encompassing most of those described in this book. As for Darfur, the IPCC's review was based on its official procedures, as well as comments and responses by experts and government representatives. The IPCC concluded that the conflict is likely to have happened regardless of climate impacts.

Linearity, changing modalities and diversity or actors and resources

This chapter has considered the main characteristics of the actor-networks resulting from the two climate-conflict debates, thereby describing key similarities and differences between the Darfur and Syria debates. It highlighted the process of mapping translations and knowledge resources for the two climate wars, and how climate-conflict links are made by articulating linear causal mechanisms.

The chapter also outlined how both climate-conflict debates evolve through three stages: narratives, technical resources to criticize or consolidate climate-conflict links, and reviews of evidence. The evolving nature of epistemic resources through the stages of translation illustrated how the types of resources and the way they are used changes, moving from fragmented and anecdotal evidence toward the involvement of technical inscription devices, and concluding with a synthesis of evidence conducted under the rules inscribed in IPCC's internationally agreed modalities.

Finally, the chapter summarized how the climate-conflict actor-networks encompass a wide range of actors ranging from UN Secretary-General through President Biden to GWR. They involve very diverse knowledge resources – in some instances the evidence is a 20-year old interview with a near-blind Darfurian sheikh, in others it is provided by 16 climate models of the CMIP5 project. And they circulate through many platforms, including newspapers, speeches, peer-reviewed articles, blogs, documentaries and YouTube videos. Darfur and Syria share the logic of the linear causal mechanisms, the evolving modalities of translation throughout the three stages of the debates, as well as many of the technical resources integrated into the translation. However, they differ in terms of the components of the mechanisms, the amount of resources mobilized, the modalities of circulation, and the way in which climate-Syria links were also supported by peer-reviewed studies.

Chapters 5 to 7 have exemplified how, in the formation of the climate-conflict nexus, each translation is itself an results of several knowledge resources. The contributions analyzed directly in this book provide only the tip of the iceberg, with each translation involving resources circulating via inscription devices in several formats, which transport information between actors and turn the world into "useable, mobile knowledge" (Best and Walters 2013, 332).

This information on the basic characteristics climate-conflict actor-networks – the linear causal mechanisms, the changing modalities of translation, and the diversity of epistemic resources – provides a basis for the following three chapters. Those chapters will look deeper into the role of knowledge within the climate-conflict debates and provide further clarity on the contradictory logic of the broader climate-security debates (Chapter 8), the unintended marginalizing consequences of the ways in which knowledge is integrated into the climate-conflict debates (Chapter 9), and how the changing modalities of translation provide an illustration of different sources of influence resulting from the ways in which actors and resources are organized (chapter 10).

Literature

Angermayer, G.; Dinc, P. and Eklund, L. 2022: *The Syrian Climate-Migration-Conflict Nexus: An Annotated Bibliography*. Center for Advanced Middle Eastern Studies, Lund. At: www.cmes.lu.se/sites/cmes.lu.se/files/2022-03/Syrian%20Climate-Migration-Conflict%20Nexus.pdf (accessed 4 August 2023).

Best, J. and Walters, W. 2013: *Actor network theory and international relationality: Lost (and found) in translation*. In: International Political Sociology, 7(3), 332–4.

Brand, R. 2015: *ISIS vs. climate change – which kills more?* The Trews. 5 March. At: www.youtube.com/watch?v=Zrr5BvrAo-Y (accessed 15 May 2021).

Brown, I.A. 2010: *Assessing eco-scarcity as a cause of the outbreak of conflict in Darfur: A remote sensing approach*. In: International Journal of Remote Sensing, 31(10), 2513–20.

Burke, M.B.; Miguel, E.; Satyanath, S.; Dykema, J.A. and Lobell, D.B. 2009: *Warming increases the risk of civil war in Africa*. In: PNAS, 106(49), 20670–4.

Callon, M. 1986: *Some elements of a sociology of translation: Domestication of the scallops and the fishermen of St. Brieuc Bay*. In: Law, J. (Ed.): Power, Action and Belief: A New Sociology of Knowledge? London, Routledge, 196–223.

De Châtel, F. 2014: *The role of drought and climate change in the Syrian uprising: Untangling the triggers of the revolution*. In: Middle Eastern Studies, 50(4), 521–35.

Femia, F.; Sternberg, T. and Werrell, C. 2014: *Climate hazards, security, and the uprisings in Syria and Egypt*. In: Seton Hall Journal of Diplomacy and International Relations, XVI(1), 71–84.

Fischetti, M. 2015: *Climate change Hastened Syria's civil war*. Scientific American. 2 March. At: www.scientificamerican.com/article/climate-change-hastened-the-syrian-war/ (accessed 12 November 2020).

Friedman, T. 2012: *The other Arab Spring*. The New York Times. 7 April. At: www.nytimes.com/2012/04/08/opinion/sunday/friedman-the-other-arab-spring.html (accessed 27 June 2022).

Fröhlich, C. 2016: *Climate migrants as protestors? Dispelling misconceptions about global environmental change in pre-revolutionary Syria*. In: Contemporary Levant, 1(1), 38–50.

Giannini, A., Saravanan, R. and Chang, P. 2003: *Oceanic forcing of Sahel rainfall on interannual to interdecadal time scales*. In: Science, 302, 1027–30.

Gleick, P. 2014: *Water, drought, climate change, and conflict in Syria*. In: Weather, Climate and Society, 6(3), 331–40.

Gleick, P. 2017: *Climate, water, and conflict: Commentary on Selby et al. 2017*. In: Political Geography, 60, 248–50.

Gore, A. Jr. 2015: *Speech at the 2015 Climate Summit of the Americas*. 7 July. At: www.
ontario.ca/page/climate-summit-americas-retrospective (accessed 19 July 2021).

Gray, R. 2015: *Did climate change trigger the war in Syria? Severe drought may have
contributed to uprising, study reveals.* Daily Mail. 2 March. At: www.dailymail.co.uk/
sciencetech/article-2976063/Did-climate-changetrigger-war-Syria-Severe-drought-
contributed-uprising-study-reveals.html (accessed 23 November 2019).

Hendrix, C.S. 2017: *A comment on "climate change and the Syrian civil war revisited".* In:
Political Geography, 60(1), 251–2.

Hoerling, M.; Eischeid, J.; Perlwitz, J.; Quan, X., Zhang, T. and Pegion, P. 2012: *On the
increased frequency of Mediterranean drought.* In: Journal of Climate. March. At: https://
journals.ametsoc.org/doi/full/10.1175/JCLI-D-11-00296.1 (accessed 27 August 2019).

Holthaus, E. 2014: *Is climate change destabilizing Iraq?* Slate. 27 June. At: https://slate.
com/technology/2014/06/isis-water-scarcity-is-climate-change-destabilizing-iraq.html
(accessed 14 May 2021).

Homer-Dixon, T. 2007: *Cause and effect.* In: SRC Blogs – Climate and Environment, Making Sense of Sudan. 2 August. At: https://homerdixon.com/cause-effect/ (accessed 5
June 2020).

Hsiang, S.M.; Burke, M.B. and Miguel, E. 2014: *Reconciling climate-conflict meta-analyses:
Reply to Buhaug et al.* In: Climatic Change, 127, 399–405.

Ide, T. 2018: *Climate war in the Middle East? Drought, the Syrian civil war and the state of
climate-conflict research.* In: Current Climate Change Reports, 4(4), 347–54.

IPCC 2022: (Pörtner, H.-O.; Roberts, D.C.; Tignor, M.; Poloczanska, E.S.; Mintenbeck,
K.; Alegría, A.; Craig, M.; Langsdorf, S.; Löschke, S.; Möller, V.; Okem, A. and Rama,
B. (Eds.)): *Climate Change 2022: Impacts, Adaptation, and Vulnerability. Contribution
of Working Group II to the Sixth Assessment Report of the IPCC.* Cambridge University Press. In Press. At: www.ipcc.ch/report/sixth-assessment-report-working-group-ii/
(accessed 19 June 2022).

Kelley, C.P.; Mohtadi, S.; Cane, M.A.; Seager, R. and Kushnir, Y. 2015: *Climate change
in the fertile crescent and implications of the recent Syrian drought.* In: PNAS, 112(11),
3241–6.

Kevane, M. and Gray, L. 2008: *Darfur: Rainfall and Conflict.* In: Environmental Research
Letters, 3. At: https://iopscience.iop.org/article/10.1088/1748-9326/3/3/034006/pdf
(accessed 21 November 2019).

Latour, B. 2005: *Reassembling the Social – an Introduction to Actor-Network Theory.*
Oxford University Press, Oxford.

Mazo, J. 2010: *Climate Conflict.* Routledge, New York.

Obama, B. 2015: *Remarks by the President at the United States Coast Guard Academy Commencement.* 20 May. At: https://obamawhitehouse.archives.gov/the-press-
office/2015/05/20/remarks-president-united-states-coast-guard-academy-commencement
(accessed 25 November 2019).

Qiu, L. 2015: *Fact-checking the link between climate change and ISIS.* Politifact.com. 23 September. At: www.politifact.com/factchecks/2015/sep/23/martin-omalley/fact-checking-
link-between-climate-change-and-isis/ (accessed 5 August 2022).

Schwartz, P. and Randall, D. 2003: *An Abrupt Climate Change Scenario and Its Implications
for United States National Security.* Washington, DC. At: https://training.fema.gov/hiedu/
docs/crr/catastrophe%20readiness%20and%20response%20-%20appendix%202%20
-%20abrupt%20climate%20change.pdf (accessed 6 August 2021).

Selby, J. and Hoffmann, C. 2014: *Beyond scarcity: Rethinking water, climate change and
conflict in the Sudans.* In: Global Environmental Change, 29, 360–70.

Selby, J.; Dahi, O.; Fröhlich, C. and Hulme, M. 2017: *Climate change and the Syrian civil war revisited*. In: Political Geography, 60(1), 232–44.

Selby, J.; Daoust, G. and Hoffmann, C. 2022: *Divided Environments: An International Political Ecology of Climate Change, Water and Security*. Cambridge University Press, Cambridge.

Verhoeven, H. 2011: *Climate change, conflict and development in Sudan: Global neo-Malthusian narratives and local power struggles*. In: Development and Change, 42(3), 679–707.

Werrell, C.E. and Femia, F. 2012: *Syria: Climate Change, Drought and Social Unrest*. CCS. 29 February. At: https://climateandsecurity.org/2012/02/syria-climate-change-drought-and-social-unrest/ (accessed 10 November 2020).

Werrell, C.E. and Femia, F. 2017: *The Climate Factor in Syrian Instability: A Conversation Worth Continuing*. CCS. 8 September. At: https://climateandsecurity.org/2017/09/08/the-climate-factor-in-syrian-instability-a-conversation-worth-continuing/ (accessed 28 July 2022).

Werrell, C.E.; Femia, F. and Sternberg, T. 2015: *Did we see it coming?: State fragility, climate vulnerability, and the uprisings in Syria and Egypt*. In: SAIS Review of International Affairs, 35(1), 29–46.

Additional literature referenced by the contributions to the climate-conflict debates

Ababsa, M. 2015: *The end of a world: Drought and agrarian transformation in northeast Syria (2007–2010)*. In: Hinnebusch, R. and Zintl, T. (Eds.): Political Economy and International Relations: Vol. 1. Syria from Reform to Revolt. Syracuse University Press, Syracuse, 199–222.

Black, R.; Bennett, S.R.G.; Thomas, S.M. and Beddington, J.R. 2011: *Climate change: migration as adaptation*. In: Nature, 478, 447–9.

De Waal, A. 1989: *Famine that Kills: Darfur, Sudan, 1984–1985*. Clarendon, Oxford.

Eklund, L. and Thompson, D. 2017: *Differences in resources management affects drought vulnerability across the borders between Iraq, Syria, and Turkey*. In: Ecology an Society, 22.

Goldstone, J. 2002: *Population and security: How demographic change can lead to violent conflict*. In: Journal of International Affairs, 56(1), 3–22.

Kitoh, A.; Yatagai, A. and Alpert, P. 2008: *First super-high-resolution model projection that the ancient "fertile crescent" will disappear in this century*. In: Hydrological Research Letters, 2, 1–4.

Nabhan, G. 2010: *Drought drives Middle Eastern pepper farmers out of business, threatens prized heirloom chillies*. Grist.org. At: http://grist.org/food/2010-01-15-drought-drives-middle-eastern-peppers (accessed 5 August 2022).

Saleeby, S. 2012: *Sowing the seeds of dissent: Economic grievances and the Syrian social contract's unravelling*. Jadaliyya. 16 February. At: www.jadaliyya.com/pages/index/4383/sowing-the-seeds-of-dissent_economic-grievances-an (accessed 15 July 2022).

Solow, A.R. 2013: *Global warming: A call for peace on climate and conflict*. In: Nature, 97(7448), 179–80.

Young, H. and Osman, A.M.K. 2005: *Darfur: Livelihoods Under Siege*. Tufts University, Feinstein International Famine Center, Medford, MA.

8 Knowledge and the making of climate-conflict links

Based on the descriptions of the climate-conflict debates in the previous chapters, this chapter consider specifically the role of knowledge resources, in particular scientific knowledge, in climate-conflict actor-networks in light of the linear causal mechanisms, modalities of translation, and the diversity of actors, resources and circulation channels described in Chapter 7. It thereby sheds light on the contradictory co-existence of disputed evidence and accelerating securitization of climate change. The chapter describes an epistemic landscape that bears little resemblance to idealized linear science-policy relations: knowledge resources are heterogenous and fragmented; the sciences work in reaction to narratives; and science-policy boundaries are regularly blurred. It also highlights how the scientific basis of the linear causal mechanisms is imbalanced toward the natural sciences, and involves very limited resources for understanding human components of the mechanisms, leading to marginalization of impacted populations, stereotypical presentations, and risks of confirmation bias in climate-conflict research. Finally, it considers how the dynamics of knowledge leads to changing patterns of influence within the climate-conflict actor-networks, describing how influence, rather than being an attribute of actors, results from changing constellations of knowledge resources and changing modalities of translations.

Knowledge within climate-conflict actor-networks

As outlined in Chapter 1, climate-security discourses involve a contradictory logic: the securitization of the "science-driven" issue of climate change keeps advancing even though the scientific evidence for climate-security links remains contested. This indicates that research findings and policymaking might be disconnected. Chapters 5, 6 and 7 described the efforts of actors to assemble knowledge resources to debate the controversies about the role of climate change in the civil wars in Darfur and Syria. The chapters mapped how climate and the two conflicts have been translated together, the central role of linear causal mechanisms as a link between climate and the conflicts and as a structure of the debates, how modalities of translations change throughout the debate, and the diversity of knowledge resources involved in the translations. Based on the mapping, this section explores the nature of the science-policy contradiction by considering specifically how

DOI: 10.4324/9781003451525-8

knowledge resources are integrated into the controversies about climate-Darfur and climate-Syria links, and, in particular, how scientific resources feature as part of the multiplicity of knowledge resources.

As outlined in Chapter 7, the knowledge resources changed significantly throughout the debate. While the initial narratives are based on diverse and fragmented resources, technical resources are mobilized to criticize or consolidate climate-conflict links, and the modalities of an IPCC review guide the synthesis of the status of climate-conflict research.

When considering the role of knowledge resources in linking climate change and the two conflicts, four key features stand out:

* Some aspects of climate-conflict debates involve hardly any visible resources.
* As a rule, scientific knowledge is integrated to the debate in reaction to initial narratives.
* Knowledge resources are often diverse and fragmented.
* Boundaries between scientific knowledge and policymaking are regularly blurred.

Connecting climate and conflicts with limited evidence

The first notable feature of the role of knowledge resources in the climate-conflict actor-networks is that several aspects of the debate involve almost no visible resources at all, in particular in the early stages. Casual references to, for example, "many studies", "UN statistics" or "scientific findings" are common.

In the case of Darfur, the last component of the linear causal mechanism – that farmers and pastoralists were driven to conflict by resource scarcity – seems to have been built on only two resources: first, an interview with Sheikh Abdallah conducted in 1985 by Alexander De Waal for his PhD fieldwork, and cited by Faris in 2007 for his Article in *The Atlantic*. Second, a statement by an anonymous North African diplomat referenced by Baldauf in his article in *Christian Science Monitor*. Other contributions that mention the farmer-pastoralist links do not identify any specific knowledge resources. Thus, social dynamics of a region of 6 million inhabitants are represented within the debate on the basis of only these two statements.

A similar case in point is the debate on climate-ISIS links. Those links were built in particular on two types of resources: actors cited in particular Hoerling et al.'s study on the severity of Mediterranean drought, and unspecified "recent studies" about correlations between heat and violence. On that basis, it was argued that perhaps the unusually hot year of 2014 and the rise of ISIS were connected. The other approach to connect the two was the notion that ISIS recruitment was particularly successful due to drought-induced poverty. For this, the only visible evidence was an interview with one tribal sheikh in northern Iraq.

A third example is how some of the specific but significant points within the debates are built on limited evidence. Chapter 6 described how the notion that 1,5 million had to flee their homes after Syria's prewar drought became a widely circulating central figure in the debates. However, the chapter also highlighted how that number originated in a statement given by one Syrian official to IRIN, which, although contradicted by other Syrian government sources around the same time, was cited by several contributions to the climate-Syria debate, took on a life on its own, and continues to be cited today.

This way, it appears that several aspects of the climate-conflict links have been built with very limited information and that dramatic statements can circulate widely without a clear and uncontested knowledge basis. In contrast, other components of the linear causal mechanisms involved much more evidence, leading to information imbalances (which will be discussed further in the chapter).

The sciences: mobilizations of technical networks to consider existing narratives

A second feature is how scientific knowledge follows the initial narratives of linear causal mechanisms, both chronologically and substantively. As illustrated by the stages of translation outlined in Chapter 7, the initial narratives are followed by articles in peer-reviewed journals to support or criticize the components of the mechanisms by mobilizing technical resources. While the first statements about climate-Darfur links were made in 2004, peer-reviewed contributions followed from 2008 onward, and the IPCCs review ten years later in 2014. Similarly, the climate-Syria debate began in 2012, the first peer-reviewed inputs to the debate were published in 2014, and the IPCC review was published in 2022.

Thus, scientific studies specifically addressing the climate-conflict links appeared *after* initial translations had already connected climate and the conflicts through the initial narratives and thus "set the agenda". Scientific resources entered the controversies after translations had already been made by others, rather than as driving forces. In addition, for both conflicts, the scientific studies also followed the initial narratives substantively. This is because they focused their support or criticisms on the components of the linear causal mechanisms articulated in the early stage of the debates. In the case of Darfur, the studies criticized assumptions about failed rains, resource scarcity and the centrality of farmer-pastoralist relations – all aspects introduced in the early stages. On Syria, the studies debated drought severity, drought-climate links, migration levels and role of migrants in protests. This way, the themes and areas of focus of the scientific studies were set by the existing narrative of linear causal mechanisms. Thus, the policy debate drive the structure of the research efforts, rather than science driving policy by identifying specific problems for policymaking.

This illustrates how scientific work in the climate-conflict debates was reactive, both temporally and in terms of substantive focus, rather than a driving force, and was about checking or consolidating the initial translations of climate-conflict links. Chapter 4 discussed the "common sense" perspective that considers science and policy as separate domains and envisages a linear relationship between the two (Weingart 1999; Jasanoff 2004; Latour 2005; Boswell 2009). Along these lines, securitization research has considered that securitization of environmental issues depends greatly on the extent to which policy actors embrace the scientific agenda

(Buzan et al. 1998). However, the climate-conflict debates paint a different picture of the role of the sciences in the securitization of climate change: there is no linear relationship between the two, and it appears that the scientific research embraced a policy agenda in this case. This way, the two evolved through mutual interaction and co-production (Jasanoff 2004).

Knowledge and linear causal mechanisms: diversity and fragmentation

Where evidence was available, it was often diverse and fragmented. Chapters 5, 6 and 7 have described how, for both conflicts, the initial narratives of linear causal mechanisms were articulated on the basis of a wide range of knowledge resources. The early Darfur narratives were built in particular on interview statements, anecdotes and the Pentagon's exploratory report on upcoming resource wars. Similar heterogeneity was visible in the initial narratives of climate-Syria links, which drew information from blogs, online journals, interviews, UN reports, as well as a cable from the US Embassy in Damascus leaked through Wikileaks. Initially, scientific research was integrated into the climate-conflict debates by referencing studies that addressed climate phenomena relating to components of the assumed linear causal mechanisms (Giannini et al. 2003 for Hoerling et al. 2012 for Syria). However, those studies did not consider climate-conflict links specifically, or refer to the ongoing conflicts in the respective regions. Their focus was on regional climate impacts. This way, whichever the original objectives of the two studies, their findings were translated by others to support the initial narratives of linear causal mechanisms alongside the information from interviews and other sources.

This heterogeneity indicates that the mobilization of knowledge resources did not follow any common standard or criteria, or an aspiration to establish a broad knowledge basis for the entirety of the linear causal mechanisms. Rather, the dynamic seems to have been about choosing what information was available at the time, and what non-climate-conflict-specific information could potentially be translated to serve as evidence within the climate-conflict debates.

This fragmentation is further indicated by the pattern of mobilizing limited evidence for some components of the mechanisms described above. The result of this pattern is that the linear causal mechanisms are greatly imbalanced in terms of their evidence basis: while components related to natural phenomena such as rainfall, vegetation levels, drought severity and drought-climate links were debated by mobilizing parts of the "vast machine" of climate sciences (Edwards 2013), the evidence involved in debating the human aspects of climate-conflict links appears curiously thin and lacks a technical basis comparable to analysis of the natural phenomena. This imbalance will be elaborated further in the next section.

So in both Darfur and Syria cases, the linking of climate change and the conflict begun with compelling and coherent "storylines" based on interviews, anecdotes and limited referencing of a small selection of scientific materials brought into the climate-conflict context, rather than with comprehensive investigations of climate-conflict links and all the components of the linear causal mechanisms. In addition, as discussed earlier, some components of the mechanism were presented without any evidence. Thus, rather than being science-driven, the initial narratives consisted of coherent storylines supported with anecdotal and loosely related evidence. In addition, several aspects of the linear causal mechanisms were based on limited, anonymous or unspecified sources. Scientific resources that address specific questions related to climate-conflict links were integrated into the debate at later stages.

Working on the science-policy boundary

As discussed in Chapter 4, one area of STS research is known as boundary work. This involves studying what happens on and across the perceived boundary between science and policymaking, and how that boundary is being constantly delineated and rearticulated through the work of actors (Jasanoff 2004; Irwin 2008). Several aspects of the climate-conflict debate involve such crossing of boundaries, illustrating their permeability and dynamic character.

For instance, some of the peer-reviewed contributions to the climate-Syria debate (e.g., Femia et al. 2014; Werrell et al. 2015) were published by authors associated with the think tank CCS. This way, a science-policy boundary is crossed by individuals involved in both research and policy work. This is an obvious dynamic given their expertise, but it does illustrate the permeability of the science-policy boundary. Another type of crossing of boundaries occurs when scientists provide inputs to the debate beyond the formal scientific channels of peer-reviewed articles, academic books and scientific conferences. Within the climate-conflict debates, this involved interview statements, contributions in blogs and media articles.

A further illustration of the crossing of boundaries within the climate-conflict debates are the "knowledge infrastructures" that enable the production and circulation of climate data. As highlighted in Chapter 7, these infrastructures, including weather stations networks, satellites, climate databases and climate model simulations, are often established and operated by government organizations such as Deutscher Wetterdienst (DWD), NASA, KNMI, NOAA, SMA and MAAR.

Finally, the IPCCs review process involves various overlaps between the traditionally assumed independent spheres of science and policy. The procedures of the IPCC review (IPCC 2013) have been adopted by national governments, and those procedures determine what qualifies as evidence. The IPCC reports involve a process of expert and government commentary and authors responses, and its final reports are adopted in plenary session involving government delegates (see Chapter 5 for the details of this process). This way, the "domain" of policy is intricately interwoven with the processes of determining the facts (see also Miller 2004; Edwards 2013).

The epistemic landscape of climate-conflict actor-networks

This section has described an epistemic landscape that bears little resemblance to idealized linear science-policy relations discussed in Chapter 4: several aspects of the climate-conflict links involved limited knowledge resources, scientific knowledge was integrated into the debates reactively after the linear causal mechanisms had been articulated and studies were organized around the structures provided by those mechanisms. This indicates that the sciences, rather than driving the debate, tend to work reactively and on the basis of existing non-science-based narratives. The knowledge resources were, in particular in the initial stages, diverse and fragmented, involving the mobilization of a mixture of anecdotal evidence, interview statements, and recontextualization of existing climate studies to fit the climate-conflict narratives. This indicates a tendency to work with what evidence happens to fit the linear causal mechanisms, rather than follow standards of evidence. In addition, the boundaries between science and policymaking were blurred in several ways. As a result, as Selby et al. have observed:

> Historically, . . . public and policy discourse on the security and geopolitical implications of climate change has been well ahead of, and often at variance with, the available scientific evidence.
>
> (Selby et al. 2017, 233)

These complexities raise several questions about the science drivenness of climate security discourses and about the role of the sciences in policy debates in general. However, the imbalances of evidence across the components of the linear causal mechanisms also appear to have further unintended consequences, which will be discussed in the next section.

Knowledge, marginalization and the risk of contamination of climate-conflict research

As discussed in the previous section and in Chapter 7, the different parts of the linear causal mechanism are supported with very different amounts and types of knowledge resources. In particular, limited evidence was mobilized to support the human components of the mechanisms – which are essential for connecting environmental developments with conflict outbreak. Sweeping statements about the social dynamics of millions of people affected by conflicts were represented based on only anecdotal evidence. This section will discuss how these imbalances illustrate blind spots in the "vast machine" of climate sciences, reproduce marginalizing tendencies within the climate-conflict debates, and may create difficulties for the integrity of scientific work.

Linear causal mechanisms and epistemic blind spots

As illustrated by Chapters 5 to 7 and in the previous chapter, there are wide discrepancies between the types and amounts of knowledge resources that inform the different components of the linear mechanisms, with significantly fewer resources mobilized on the components involving human interactions. Throughout these stages of translations, information imbalances emerged across the components of the mechanisms. While some parts of the controversies (e.g. whether rains failed in Darfur, the drought severity in Syria, and whether Syria's drought was connected with climate change) involved mobilizing many detailed technical resources, others (including farmer-pastoralist tensions, the role of migrants in Syria's 2011 protests, the notions of causality under which climate be considered a conflict cause, or whether climate change facilitated the rise of ISIS) involved knowledge resources that tended to be fragmented and anecdotal.

The imbalances showed a similar pattern across the two conflicts: technical inscription devices featured strongly in controversies about climatic phenomena, and much less so when it comes to human and social aspects of climate-conflict links. Significant data arrangements and methodological tools were available for measuring rainfall and drought levels, conducting attribution studies, and making projections about the climate. In contrast, tools of comparable scope and efficiency seem to be unavailable for addressing the human dimensions of complex situations like the two conflicts. This illustrates how the "vast machine" (Edwards 2013) of climate sciences – a global network for understanding the climate – sheds light on only a limited part of the mechanisms linking climate and conflicts. The machine offers many tools to quantify physical phenomena, but few to represent human interactions, and, though conflict datasets and models are developing further, nothing comparable exists when it comes to understanding human interactions.

In the case of Syria, this imbalance has been noted by several critics. Selby et al. (2017) criticized the lack of references to conflict literature and ethnographic data in the climate-Syria debate. Angemayer et al.'s review of the climate-Syria studies observed that those describe migration in a static way, and do not reflect the dynamic and multi-directional nature of migration, engage with migration literature, or pay attention to post-migration livelihoods and experiences of Syrians (2022, 13). In addition, Ide highlighted how, on the one hand, information on motivations of migrants is limited, and, on the other, interview-based studies have been undervalued by quantitative researchers (2018, 353)

In the case of climate-conflict debates, the imbalances are possibly due to a supply bias: the comparable availability of tools to measure rainfall and other climate parameters in contrast to monitoring patterns of organized violence and economic activity in remote locations might make it inevitable that climate data arrangements are more advanced, and, when debating climate-conflict links, supply drives the scope of the analyses. Another driver of the information imbalances might relate to the need to articulate coherent linear causal narratives. Such narratives are needed to connect climate and the conflicts in a public discourse. But the narratives also need to be kept intact – if one piece is removed, the entire causal link collapses. Thus, the narrative coherence of the linear causal mechanisms requires including some components even if no robust evidence is available to support them.

Local populations: avatars of resource scarcity?

Whatever the drivers of the information imbalances across the components of the linear causal mechanisms, they have the consequence of reproducing patterns of marginalization of people caught in the conflicts and consolidating stereotypical assumptions about local populations. The relative absence of Darfurians and Syrians from the climate-conflict debate raises questions about the basis with which the human components of the linear causal mechanisms are built in the end.

Very few voices of those impacted by the conflicts are heard in the debates. In the case of Darfur, the only visible resources to support statements about the role of farmer-pastoralist tensions in the 2003 conflict were a statement by a North African diplomat, and a 1985 interview with a Darfurian Sheikh. There are simply no other visible resources to demonstrate the last component of the climate-Darfur mechanism. While the preparation of UNEP's (2007) report involved interviewing two thousand Darfurians (see Chapter 5), any potential light that those interviews could shed on farmer-pastoralist relations is not specified in the report. Thus, it seems that the climate-Darfur debate involved a significant under-representation of Darfurians themselves and the information that they might be able to share about the factors that led to the conflict. Along similar lines, Selby et al. have observed that "climate-conflict accounts of the war in Darfur are consistently wanting in their characterization of local livelihoods and the region's political ecology" (2022, 72). The same applies, though perhaps less drastically, for example to statements about the role of drought migrants in Syria's 2011 protests. These statements were built on data on Syrian overall population growth, a UNHCR survey highlighting the pressures created by Iraqi refugees, and literature that connects demographic growth and instability, concluding, that is, that migration caused a "population shock" around Syria's cities (Kelley et al. 2015, 3242). However, the specific notion that migrants from the countryside had a central role in Syria's protests came from a statement by one farmer and an interview with one Syrian refugee. Among the more critical studies, further interviews with Syrian refugees were mobilized by Fröhlich (2016).

Thus, in both cases, the human components of the linear causal mechanisms are based on limited direct information from those affected, constituting a marginalization of the local populations. However, they also tend to invite stereotypical statements about resource scarcity and the risks of migration. In the context of Darfur, Verhoeven (2011) observed that, in the absence of information, Western imagination and narratives tend to treat Darfur as a canvas for assumptions of "Malthusian societal collapse" – meaning a scenario where population growth outpaces available resources and leads to violent conflict – despite the lack of evidence for such dynamics (680). The lack of adequate information on the human aspects of the linear causal mechanism seems to open a window for actors to hastily fill the gaps with stereotypical assumptions about the country or region in question. In both Darfurian and Syrian cases, local people (and governments) are portrayed as helpless victims of environmental disaster and resource scarcity with no coping possibilities or agency – portrayals that Selby and Hoffmann described, in the case of Darfur, as "outdated" (2014). A further consequence was, as observed by several contributions, that such narratives also enabled powerful actors to deny their responsibility (Salehyan 2007; Verhoeven 2011; Selby and Hoffmann 2014).

Latour (2005) emphasized that social sciences should not treat the people they study as "avatars" of social forces. But this clearly happens with Darfurians and

Syrians in the climate-conflict debates through a vicious circle of limited evidence and stereotypical assumptions: when limited information about local social dynamics is available, the incentives to populate the linear causal mechanisms with stereotypical assumptions are strong. In turn, the incentives to mobilize information about local populations are reduced because the linear causal mechanisms already explain what is happening. This way, scarcity-based climate-conflict narratives do not require human agency.

Risks of confirmation bias in climate-conflict research

In addition to the epistemic blind spots and tendencies to marginalize local populations, a further risk of the imbalanced information basis of the linear causal mechanisms is that it might incentivize a confirmation bias in climate-conflict research. This chapter has described how scientific studies tend to follow the structure of the narratives of the linear causal mechanisms. However, the need to maintain the coherence of the mechanisms possibly invites researchers to provide supporting contributions based on limited evidence.

For example, in the climate-Syria debate, some of the scientific studies supportive of climate-conflict links, in particular Kelley et al. (2015) provided supportive evidence on all components of the linear causal mechanisms, including those related to social dynamics, which, as described earlier, were backed by limited knowledge resources. It is possible that the need to maintain the coherence of the linear causal mechanism compelled some researchers to address components on which limited evidence was available, or where conclusions about conflict dynamics had to be extrapolated on the basis of associated information. This was done, for example, by using information on Syria's population growth and conflict literature to argue that migration pressures destabilized Syria's cities.

The challenge with linear causal mechanisms is that if one component is left blank, the consolidation of the mechanism would be difficult, and the proponents of climate-conflict links would need to acknowledge that major gaps remain. Then the entire mechanism would collapse, since its linear nature requires all its components to be intact. It is possible that some supporters of climate-Syria links, faced with the need to consolidate the linear causal mechanism, but also with the absence of proper information to support its human components, opted to fill those gaps with limited information and for example Malthusian narratives about population shocks and limited resources (Verhoeven 2011). Thus, the demands of maintaining the initial narratives of the linear causal mechanism might create a risk of confirmation bias for the sciences intending to fortify those narratives.

Knowledge and patterns of influence

A further dimension of the climate-conflict actor-networks are the shifting patterns of influence throughout the stages of translation. In IR, power is often conceptualized as something stable that subjects hold and practice by possessing certain resources (Mattern 2008, 692; Rosecrance 2008, 721). In contrast, Braun et al. (2019) have emphasized how IR research can benefit from approaching agency as a practical achievement of embeddedness in networks. As discussed in chapter 4, from an ANT-perspective, agency, power and influence are results of heterogeneous networks of associations between actors (Latour 2005; Best and Walters 2013; Braun et al. 2019). In Latour's words, "power" can be described a result of various processes (2005, 64), rather than a theoretically derived driving force of processes. This section will focus on how the translations and mobilization of knowledge resources in the two climate-conflict debates resulted in a constraining and enabling actor-network, which involved changing positions of agency and influence. Each of the three stages of translations identified in chapter 7 can be characterized by different patterns of influence.

In both Darfur and Syria, actors first articulated the components of the linear causal mechanisms (stage 1). The components established the structure for the debate and thus steered the work of actors that follow in stages 2 and 3 toward specific arguments and knowledge resources as they directed their contributions to criticize or support the components. In the first stage, influence appeared to reside in the capacity to articulate and circulate compelling and logical "stories" of climate-conflict links and to establish those stories as the structure of the debate, thus co-determining future contributions. When scientific resources entered the debates in stage 2, various inscription capabilities that enabled representing the climate pasts and futures of Darfur or Syria emerged as factors of agency and influence, and thus, the several new knowledge resources mobilized began to influence the dynamics of the actor-networks. Finally, in response to the increasing complexity, efforts were made to review evidence and draw conclusions, including through the official reviews by the IPCC, which was mandated to include and exclude evidence in its synthesis. This way, the IPCC procedures provided mechanisms of influence that could remove contributions that did not conform with its criteria from serving as evidence.

In a nutshell, these were the three types of influence discernible within the actor-networks: the power to articulate coherent narratives, the power to collect and represent distant places and phenomena, and the power to review, synthesize, include and exclude. These patterns shift based on the status of the debate and predominant knowledge resources of each stage, and involved dynamics where initially influential actors disappeared from the debate as the modalities of translation and knowledge resources changed.

Coherence as influence

As introduced in Chapter 7, the rather fragile components of the linear causal mechanisms in the first stages evolved into the structure for the two climate-conflict debates. The mechanisms provided coherent explanations of how climate and the conflicts were connected. In other words, they offered compelling and causally plausible, or "intuitively sensible" (Verhoeven 2011, 680) storylines for connecting the two.

Had the mechanisms not been questioned by others, they might have perhaps become consolidated as black boxes and thus unquestionable premises of climate-conflict debates. However, as outlined in Chapter 4, translations are a collective process and can always involve challenges that prevent the formation of black boxes. The mechanisms invited further contributions to join the conversation and to mobilize knowledge resources to debate the resulting controversies. As described in Chapters 5 to 7, the components of the linear causal mechanisms become the central controversies around which other actors undertake translations. This way, they influenced the contours of the rest of the debates.

More specifically, the narratives of the linear causal mechanisms influenced other actors by compelling them to organize their contributions, and to mobilize further resources, around the mechanism components. The articulations of the mechanisms exerted influence by becoming the structure of the climate-conflict debates. They influenced the focus of other actors' contributions and the choice of knowledge resources they mobilized. In the case of Darfur, the validity of the three components of the mechanism became the main target of the critical contributions that followed, which question whether the rains failed (Kevane and Gray 2008), whether there was resource scarcity (Brown 2010), and whether farmer-pastoralist assumptions are credible (Selby and Hoffmann 2014). Similarly, initial articulations of the mechanism for Syria were established as the structure on which the contributions that followed commented on (both supporting (Gleick 2014; Kelley et al. 2015) and critical ones (De Châtel 2014; Fröhlich 2016; Selby et al. 2017)).

Thus, influence was generated through the articulations in stage 1, even with limited resources, of coherent narratives that compelled others that followed to consider the components of those narratives, but that also invited new actors that could mobilize technical resources to speak for the climate and social past and future of Darfur and Syria to debate the controversies.

Technical resources as a source of influence

In the next stage, articles in peer-reviewed journals presented criticisms or supporting statements related to the components of the linear causal mechanisms, thus following their structuring influence. On the one hand, the influence of the initial narratives oriented the peer-reviewed contributions. But, on the other, the new contributions gained influence by mobilizing technical resources and by virtue of their publication in academic journals – the latter characteristic being relevant in the context of the IPCC reviews (see next section).

Influence in this stage involved the capacity to acquire data from weather stations, satellites or interviews; to draw information from climate model simulations; to apply statistical methodologies and indices; to cite literature; as well as to present outcomes in figures and maps to present the climate and social pasts and futures of Darfur and Syria. Throughout this process, the capacity to "speak for" Darfur and Syria was acquired by mobilizing complex chains of translations that included measurement devices that monitor rainfall, data processing modalities and repositories operated by institutions such as the Sudanese Meteorological Agency and the UEA CRU, distribution arrangements for datasets, and statistical methods used to convert the data into figures that represented developments and trends in Darfur and Syria or the wider regions, as well as the authority achieved through peer-review of academic journals.

For example, after Ki Moon (2007), Faris (2007) and UNEP (2007) claimed that rains failed in Darfur, Kevane and Gray (2008) mobilized rainfall data, gridded datasets, statistics, figures and tables, to argue that they did not. For their consolidation of climate-Syria links, Kelley et al. (2015) assembled precipitation data, statistical methods and drought indices to confirm drought severity; climate model projections to demonstrate climate-drought connections; Syrian government sources to estimate migration levels; and conflict literature to show the destabilizing effects of demographic change. Similarly, after Werrell and Femia (2012), Friedman (2013) and Femia et al. (2014) claimed that protests in Syrian cities were associated with drought migrants, Fröhlich (2016) conducted 32 interviews in refugee camps in Turkey and Jordan to question the claims and Selby et al. (2017) cited literature on the motivations of the protesters to suggest that the uprising were rather due to repression and lack of opportunities. This way, the components of the linear causal mechanism led to the integration of new actors, which gained agency and influence within the actor-networks. But the actors in this stage operated under the influence of the linear causal mechanism articulated in the first stage, even as they criticized it, given that their contributions were targeted toward dismantling its components.

This stage also illustrated the fragility of the linear causal mechanism: with each component providing the basis for next, each had to remain intact for the mechanism to remain in place. In this stage, those components turned into sites of conflict.

In the climate-Syria debate, the science-based consolidation changed the dynamics of influence: first, by adding a layer of evidence to back up the components of the mechanism, Kelley et al. (2015) establish a structure in which the critics of the mechanism were now confronted with the challenge of dealing with the evidence mobilized by the authors, not just with the initial narratives that involved limited resources. In the case of Darfur, the initial narrative were based on limited evidence, and were thus "easy targets" for science-based scrutiny. However, in the case of Syria, the critical contributions had to address the evidence presented by the science-based contributions. A second difference related to the IPCC review. In the case of Darfur, the IPCC excluded all initial articulations of the linear causal mechanism from its review since they did not confirm to its criteria for evidence. However, the peer-reviewed consolidations of the mechanism by Kelley et al. (2015) and others were clearly within the sphere of the IPCC's criteria, which preempted any future rejection by the IPCC. These modalities thus helped "fortify" the initially fragile linear causal mechanism against criticism and exclusions by the IPCC.

These new actors and knowledge resources also increased the complexity of the debate. In stage 1, narratives were constructed mainly on the basis of coherence and consistency with existing climate research and assumptions about resource scarcity. However, in stage 2, influence emerged from a large network of actors that delivered data from distant locations (rainfall stations in Darfur, orbiting satellites that monitor vegetation, or interviews with Syrian refugees), distributed that data, organized it into datasets that enable statements about long-term trends and apply statistical tests, and presented it in figures and tables. As the controversies intensified, a growing network was required to sustain the debate and to speak for distant people, places and climates, and consolidate positions, and this complexity incentivized the stage of review, where yet a different form of influence was discernible: the capacity to include and exclude evidence.

Inclusion and exclusion of evidence

Once the number of actors and the complexity of the debate increased, the need arose to synthesize the information in order to continue a meaningful policy debate, opening the door for another type of agency and influence: that of synthesis, inclusion and exclusion – stage three of the climate-conflict debates described in chapter 7. Under its mandate from governments, the IPCC considered the evidence for climate-conflict links for Darfur and Syria in 2014 and 2022, respectively. The review process demonstrated in particular two factors of influence: selection criteria for what qualifies as evidence, and the process of expert and government review.

The IPCC's influence, inscribed in its 2013 procedures, takes in particular the form of including and excluding evidence. The IPCC does not conduct new studies, but reviews existing ones to generate information for policymaking. Its work is guided by modalities that enforce definitions of what qualifies as scientific. As described in Chapters 5 and 7, the IPCC has been mandated by the governments to synthesize scientific research on climate change, and it applies an internationally agreed typology and criteria to select what types of publications qualify for its review. The criteria include peer-reviewed literature; reports by governmental, industrial, research and international organizations; and conference proceedings. But they exclude journalistic and media resources, blogs, social media and personal communications (IPCC 2013, 17). By applying the criteria, the IPCC exerts influence by limiting the types of information that is considered in its synthesis. In the case of the two climate-conflict debates, this means that all contributions that do not meet the criteria, including all of the contributions involved in the initial articulations of the linear causal mechanisms (which were published mainly in media contributions, blogs, speeches and online platforms) were not considered in the reviews. For example, speeches by Presidents Sarkozy and Obama and various government ministers, and UN Secretary-General Ban Ki Moon's *Washington Post* article, are not considered. None of the many newspaper and media articles and reports by NGOs are considered in the reviews. Debates at the UNSC do not make the cut. Thus, a majority of the climate-conflict debates is "erased" as it does not meet the inclusion criteria.

It is notable that these government-defined criteria explicitly excluded media, non-governmental and online resources. In the context of conflicts, much of the latest information is generated by journalists and non-governmental organizations operating in the frontlines of crises in territories inaccessible to scientific studies. They often create information at a much shorter turnaround time than the long processes of research and peer-review. In addition, online resources are becoming increasingly common means of communication today.

For Darfur this meant that all of the contributions in the first stage, including those by Ban Ki Moon, UK ministers, UNEP and many newspapers, were not considered. The actors that were influential in launching the debate in 2006–2008 by articulating the linear causal mechanism and thus determining the structure of the debate were categorized into relative irrelevance by the IPCC's criteria. In contrast, the criteria also meant that all peer-reviewed articles were considered, and, as all of those were critical of climate-Darfur links, the IPCC review was essentially a synthesis of the criticisms of climate-Darfur links. In the case of Syria, Chapters 6 and 7 described how several peer-reviewed studies mobilized evidence to consolidate the components of the linear causal mechanisms. This means that, in contrast to Darfur, several scientific studies supportive of climate-Syria links were in the sphere of the IPCC review. Subsequently, contributions supportive of climate-Syria links ended up in comparably more influential positions than in the case of Darfur.

Here, agency and influence emerged from the mandate to synthesize for policymakers, from the criteria to select which statement qualify as evidence, and through the modality of expert and government review. The review enabled in particular researchers and government delegates to influence the outcome of the synthesis. This influence changed the status of all previous statements, either increasing their credibility as evidence, or discarding them as evidence. Thus, the IPCC process took parts of the past debate, labeled those as acceptable evidence, and left the rest out of its consideration.

However, regardless of the "significant power" given to the IPCC through the legitimation by governments (Miller 2004) and its formal function of establishing facts, it is not clear whether its rather critical conclusions of climate-conflict links for both Darfur and Syria really have provided the last word in the two debates. Since the IPCC's 2014 review of climate-Darfur links, Darfur has regularly been mentioned as an example for climate-conflict links (e.g. Adelphi 2015; GWR 2019; The Economist 2019; Biden 2020; World Food Program 2020). On the one hand, while the IPCC's review of climate-Syria links was published in 2022, and limited further contributions to the debate after it were available for this book, at least some have continued to characterize climate change a key factor in the Syrian conflict (MedGlobal 2022). On the other, it seems that the Darfur and Syria conflicts have received less public attention as examples of climate wars as in the past, and the debate has focused on a broader range of conflicts and questions of environmental peacebuilding, and been less about framing existing major conflicts as climate wars. But it remains to be seen whether the IPCCs reviews, with all their authority, can provide clear and conclusive answers to the climate-conflict controversies, or whether the contradictory co-existence of limited evidence and accelerating securitization will continue in the future.

Knowledge and the climate wars

This chapter described three key modes of ordering of climate-conflict actor-networks related to the role of knowledge. First, what do the ways in which knowledge resources have been integrated into the debate say about their role in the climate-conflict debates. Second, how the imbalances of information across the components of the linear causal mechanisms illustrate epistemic blind spots of global knowledge arrangements, reproduce patterns of marginalization of impacted populations, and may create risks of confirmation biases in climate-conflict research. And third, how the changing modalities of translation throughout the debate illustrate evolving patterns of influence within the debate.

It was highlighted how several aspects of the climate-conflict links – especially those related to social dynamics in the impacted countries – involved very limited knowledge resources, illustrating how statements within the narratives of linear causal mechanisms can be made with a limited knowledge basis. Scientific knowledge was generally integrated into the debate after the linear causal mechanisms had been articulated, and organized around the structures provided by the initial narratives, indicating that scientific inputs do not drive the debate, but rather work reactively to scrutinize the debate. Knowledge resources in the initial stages are often diverse and fragmented and do not seem to follow any common criteria, except perhaps the focus on finding the evidence that fits or can be made to fit the narratives of the linear causal mechanisms. In addition, the boundaries between science and policymaking are blurred as several actors work on both sides of the assumed science-policy boundary, governments operate knowledge arrangements, and the IPCC involves an interplay between government and scientific actors.

This chapter also discussed how the ease with which linear causal mechanisms were articulated and the fragmented character of knowledge resources resulted in significant information imbalances between nonhuman and human aspects of the climate-conflict debates. While the "natural" parts of the linear causal mechanisms were sustained with scientific evidence, the human aspects were generally introduced and reproduced with anecdotal and very limited information. This exemplified how the "vast machine" of the climate sciences involves many tools for understanding natural phenomena but no comparable tools are available for understanding human components of climate-conflict links. As many advocates of climate-conflict links have focused on defending the coherence of the linear causal mechanisms regardless of these epistemic blind spots, there has been a tendency to introduce stereotypical and outdated assumptions about local populations to fill the information gaps, to reproduce patterns of marginalization of people caught in the conflicts, and possibly to create risks of confirmation bias as scientific studies have sought to confirm the components of the linear causal mechanisms.

In addition, the implications of the changing patterns of influence within the climate-conflict actor-networks were discussed. The chapter highlighted the factors of influence within complex science-policy debates, in particular the capacity to articulate and circulate coherent narratives about climate-conflict links, the ability to mobilize technical resources and the official and internationally codified power to review, include and exclude evidence. It described how, rather than some actors wielding power due to their individual characteristics, influence is a result of a bundling of the types of knowledge resources circulated by the actors, and the inscribed rules for the consideration of those materials. Thus, influence takes different forms, resides in multiple actors, and depends on the existing translation modalities and the inscription devices involved in the process. However, whether even the IPCC reviews have the capacity to conclude climate-conflict debates remains to be seen. Finally, it was highlighted that while these dynamics of influence are essential features of the scientific correction mechanisms that help enhance the epistemic basis of policy debates, they also might marginalize diverse forms of knowledge – in particular journalistic and non-governmental sources – that could account for local circumstances in conflict zones.

How can these considerations help us understand the contradictory co-existence of disputed evidence and accelerating securitization of climate change – usually considered and expected to be a very "science-driven" issue? It appears that any perception of contradiction depends on assuming that there is a clear "firewall" between the sciences and policymaking. If one considers that there is a "scientific agenda" and a "policy agenda" (Buzan et al. 1998), where scientific knowledge identifies issues and solutions, and then policymakers take decisions about possible responses, then the dynamics of the securitization of climate change might seem contradictory as the domain of science cannot agree on the issues. However, this book indicates that the two climate-conflict debates were first of all a result of the narrative coherence of the linear causal mechanisms, and only later a subject of scientific disputes. The dynamics of knowledge within the climate-conflict debates are much more complex than simply "science-driven", and they involve many instances of blurred lines between science and policymaking. In light of this complexity, there is no contradiction. However, this does not mean that things are quite satisfactory, as the complex role of knowledge is likely to reduce the quality of policymaking, provides entry points for climate skeptics to question the entire climate policy agenda and creates patterns of marginalization and risks of confirmation biases. However, as discussed in this section, power and influence are dynamic and can change with the evolving modalities of translation, indicating that changes are possible. The next chapter will discuss how the knowledge basis of climate-security and other complex international debates could be brought on a stronger epistemic footing.

Literature

Adelphi 2015: *Drought, Migration and Civil War in Darfur (ECC Factbook Conflict Analysis)*. Video published on YouTube on 28 October. At: www.youtube.com/watch?v=_MF2ZAHDdoQ (accessed 5 March 2021).

Angermayer, G.; Dinc, P. and Eklund, L. 2022: *The Syrian Climate-Migration-Conflict Nexus: An Annotated Bibliography*. Center for Advanced Middle Eastern Studies, Lund. At: www.cmes.lu.se/sites/cmes.lu.se/files/2022-03/Syrian%20Climate-Migration-Conflict%20Nexus.pdf (accessed 4 August 2023).

Best, J. and Walters, W. 2013: *Actor-network theory and international relationality: Lost (and found) in translation*. In: International Political Sociology, 7(3), 332–4.

Biden, J. Jr. 2020: *Statement at the CNN Democratic Presidential Debate*. 15 March. At: www.youtube.com/watch?v=2z613_M5gxE (accessed 9 November 2020).

Boswell, C. 2009: *The Political Uses of Expert Knowledge: Immigration Policy and Social Research*. Cambridge University Press, Cambridge.

Braun, B.; Schindler, S. and Wille, T. 2019: *Rethinking agency in international relations: Performativity, performances and actor-networks*. In: Journal of International Relations and Development, 22, 787–807.

Brown, I.A. 2010: *Assessing eco-scarcity as a cause of the outbreak of conflict in Darfur: A remote sensing approach*. In: International Journal of Remote Sensing, 31(10), 2513–20.

Buzan, B.; Waever, O. and de Wilde, J. 1998: *Security – A New Framework for Analysis*. Lynne Rienner Publishers Inc., London.

Callon, M. 1986: *Some elements of a sociology of translation: Domestication of the scallops and the fishermen of St. Brieuc Bay*. In: Law, J. (Ed.): Power, Action and Belief: A New Sociology of Knowledge? London, Routledge, 196–223.

De Châtel, F. 2014: *The role of drought and climate change in the Syrian uprising: Untangling the triggers of the revolution*. In: Middle Eastern Studies, 50(4), 521–35.

The Economist 2019: *How climate change can fuel wars*. The Economist. 23 May. At: www.economist.com/international/2019/05/23/how-climate-change-can-fuel-wars (accessed 23 May 2023).

Edwards, P.N. 2013: *A Vast Machine – Computer Models, Climate Data, and the Politics of Global Warming*. MIT Press, Cambridge, MA.

Faris, S. 2007: *The real roots of Darfur*. Atlantic Monthly. April. At: www.theatlantic.com/magazine/archive/2007/04/the-real-roots-of-darfur/305701/ (accessed 21 October 2019).

Femia, F.; Sternberg, T. and Werrell, C. 2014: *Climate hazards, security, and the uprisings in Syria and Egypt*. In: Seton Hall Journal of Diplomacy and International Relations, XVI(1), 71–84.

Friedman, T. 2013: *Without water, revolution*. The New York Times. 18 May. At: www.nytimes.com/2013/05/19/opinion/sunday/friedman-without-water-revolution.html (accessed 23 November 2019).

Fröhlich, C. 2016: *Climate migrants as protestors? Dispelling misconceptions about global environmental change in pre-revolutionary Syria*. In: Contemporary Levant, 1(1), 38–50.

Giannini, A.; Saravanan, R. and Chang, P. 2003: *Oceanic forcing of Sahel rainfall on inter-annual to interdecadal time scales*. In: Science, 302, 1027–30.

Gleick, P. 2014: *Water, drought, climate change, and conflict in Syria*. In: Weather, climate and society, 6(3), 331–40.

GWR 2019: *First Climate Change War*. At: www.guinnessworldrecords.com/world-records/first-climate-change-war (accessed 7 September 2021).

Hoerling, M.; Eischeid, J.; Perlwitz, J.; Quan, X.; Zhang, T. and Pegion, P. 2012: *On the increased frequency of Mediterranean drought*. In: Journal of Climate. March. At: https://journals.ametsoc.org/doi/full/10.1175/JCLI-D-11-00296.1 (accessed 27 August 2019).

Ide, T. 2018: *Climate war in the Middle East? Drought, the Syrian civil war and the state of climate-conflict research*. In: Current Climate Change Reports, 4(4), 347–54.

IPCC 2013: *Procedures for the Preparation, Review, Acceptance, Adoption, Approval and Publication of IPCC Reports*. At: https://archive.ipcc.ch/pdf/ipcc-principles/ipcc-principles-appendix-a-final.pdf (accessed 5 June 2020).

IPCC 2014 (Adger, W.N.; Pulhin, J.M.; Barnett, J.; Dabelko, G.D.; Hovelsrud, G.K.; Levy, M.; Oswald Spring, Ú. and Vogel, C.H.): *Human security*. In: Field, C.B.; Barros, V.R.; Dokken,

D.J.; Mach, K.J.; Mastrandrea, M.D.; Bilir, T.E.; Chatterjee, M.; Ebi, K.L.; Estrada, Y.O.; Genova, R.C.; Girma, B.; Kissel, E.S.; Levy, A.N.; MacCracken, S.; Mastrandrea, P.R. and White, L.L. (Eds.): Climate Change 2014: Impacts, Adaptation, and Vulnerability. Part A: Global and Sectoral Aspects. Contribution of Working Group II to the Fifth Assessment Report of the IPCC. Cambridge University Press, Cambridge and New York, NY, 755–91.

Irwin, A. 2008: *STS perspectives on scientific governance*. In: Hackett et al. 2008, 583–608.

Jasanoff, S. (Ed.) 2004: *States of Knowledge – The Co-Production of Science and Social Order*. Routledge, Oxon.

Kelley, C.P.; Mohtadi, S.; Cane, M.A.; Seager, R. and Kushnir, Y. 2015: *Climate change in the fertile crescent and implications of the recent Syrian drought*. In: PNAS, 112(11), 3241–6.

Kevane, M. and Gray, L. 2008: *Darfur: Rainfall and conflict*. In: Environmental Research Letters, 3. At: https://iopscience.iop.org/article/10.1088/1748-9326/3/3/034006/pdf (accessed 21 November 2019).

Ki Moon, B. 2007: *A climate culprit in Darfur*. Washington Post. 16 June. At: www.washingtonpost.com/wp-dyn/content/article/2007/06/15/AR2007061501857.html (accessed 21 October 2019).

Latour, B. 1987: *Science in Action – How to Follow Scientists and Engineers Through Society*. Harvard University Press, Cambridge, MA.

Latour, B. 2005: *Reassembling the Social – an Introduction to Actor-Network Theory*. Oxford University Press, Oxford.

Mattern, J. 2008: *The concept of power and the (un)discipline of international relations*. In: Reus-Smit, C. and Snidal, D. (Eds.) 2008, 691–8.

Mazo, J. 2010: *Climate Conflict*. Routledge, New York.

MedGlobal 2022: *Climate Change, War, Displacement, and Health: The Impact on Syrian Refugee Camps*. 20 September. At: https://reliefweb.int/report/syrian-arab-republic/climate-change-war-displacement-and-health-impact-syrian-refugee-camps (accessed 13 May 2023).

Miller, C. 2004: *Climate science and the making of a global political order*. In: Jasanoff, S. (Ed.) 2004: States of Knowledge – The Co-Production of Science and Social Order. Routledge, Oxon, 46–66.

Reus-Smit, C. and Snidal, D. (Eds.) 2008: *Oxford Handbook of International Relations*. Oxford University Press, Oxford.

Rosecrance, R. 2008: *The failure of static and the need for dynamic approaches to international relations*. In: Reus-Smit, C. and Snidal, D. (Eds.) 2008, 716–24.

Salehyan, I. 2007: *The new myth about climate change*. In: Foreign Policy. 14 August. At: https://foreignpolicy.com/2007/08/14/the-new-myth-about-climate-change/ (accessed 3 November 2020).

Selby, J.; Dahi, O.; Fröhlich, C. and Hulme, M. 2017: *Climate change and the Syrian civil war revisited*. In: Political Geography, 60(1), 232–44.

Selby, J.; Daoust, G. and Hoffmann, C. 2022: *Divided Environments: An International Political Ecology of Climate Change, Water and Security*. Cambridge University Press, Cambridge.

Selby, J. and Hoffmann, C. 2014: *Beyond scarcity: Rethinking water, climate change and conflict in the Sudans*. In: Global Environmental Change, 29, 360–70.

UNEP 2007: *Sudan – Post-Conflict Environmental Assessment*. At: https://postconflict.unep.ch/publications/UNEP_Sudan.pdf (accessed 21 November 2019).

Verhoeven, H. 2011: *Climate change, conflict and development in Sudan: Global neo-Malthusian narratives and local power struggles*. In: Development and Change, 42(3), 679–707.

Weingart, P. 1999: *Scientific expertise and political accountability: Paradoxes of science in politics*. In: Science and Public Policy, 26(3), 151–61.

Werrell, C.E. and Femia, F. 2012: *Syria: Climate change, drought and social unrest*. CCS. 29 February. At: https://climateandsecurity.org/2012/02/syria-climate-change-drought-and-social-unrest/ (accessed 10 November 2020).

Werrell, C.E.; Femia, F. and Sternberg, T. 2015: *Did we see it coming?: State fragility, climate vulnerability, and the uprisings in Syria and Egypt*. In: SAIS Review of International Affairs, 35(1), 29–46.

World Food Program USA 2020: The First Climate Change Conflict. 19 December. At: www.wfpusa.org/articles/the-first-climate-change-conflict/ (accessed 13 May 2023).

9 Finding a balance between knowledge and narratives

When Prime Minister Johnson chaired the UNSC debate on climate change on 23 February 2021, he introduced climate-security links in his usual colorful fashion:

> It is absolutely clear that climate change is a threat to our collective security and the security of our nations. And I know there are people around the world who will say this is all kind of "green stuff" from a bunch of tree-hugging tofu munchers and not suited to international diplomacy and international politics. I couldn't disagree more profoundly.
>
> (UNSC 2021)

As described in this book, a growing number of actors agree with the former prime minister. Others disagree, leading to controversies, in particular about the role of climate in specific conflicts. Yet even as the evidence remains disputed, efforts to link climate and conflicts are intensifying. Chapter 2 illustrated how climate-security links are debated by policymakers, scientists, journalists, humanitarian workers, entertainers, and even the GWR.

The world expects the response to climate change to be based on the "best available science" (Paris Agreement, preamble). However, as outlined in Chapter 2, the co-existence of science-drivenness of climate issues, limited evidence for climate-security links, as well as intensified securitization, seems contradictory. If the world aspires toward science-driven climate policies, then why is climate change being increasingly securitized regardless of contested evidence? To shed light on this contradiction, this book explored how science-driven the formation of the climate-security nexus is by mapping the role of knowledge resources in the climate-conflict debates on the two "climate wars" of Darfur and Syria, both of which have been key drivers in efforts to link climate and security.

This chapter recaps the approach and key findings of this book. It then highlights the challenge of knowledge and nexus formation in a post-truth era, and discusses possible solutions to it. It describes how the fragmentary integration of the sciences in climate-conflict debates and the focus on narrative coherence contradicts global aspirations toward science-based policies, leads to the marginalization of affected populations, risks promoting suboptimal policymaking, and fuels climate skepticism. Regardless of the challenges, the chapter argues that against

DOI: 10.4324/9781003451525-9

the background of dynamic nature of power and influence described in this study, inscribed knowledge arrangements, institutions and rules of engagement are not cast in stone, but, with balanced adjustments, provide multiple channels toward change. The chapter then discusses how solutions could involve more interactive research modalities, arrangements to integrate information from the front lines of conflicts while ensuring its quality, and by using existing institutions as a basis for correcting information imbalances and for addressing the continuum from environmental change to social impacts and thus help articulate scientifically informed nexus policies. Finally, the chapter identifies potential areas for future research to enhance understanding of nexus formation in IR research.

Approach of this book

While the domains of climate and security have their own practices, discourses, institutions and actors, efforts to merge them have intensified in the past 20 years. Studying this merging is important because the two are arguably the most central issues in contemporary international politics, have a significant influence on other issues and involve calls for extraordinary measures. As illustrated in Chapter 2, quantitative correlations between climate and conflicts have been extensively studied, and climate-security links have been considered through securitization and discourse analytical approaches. The formation of issue-linkages and nexuses between issues of international politics has been researched in particular by Ernst Haas and development studies experts. Haas (1980) has also recognized the central role of knowledge in linking issues of international politics.

To complement existing research, this book offered a new perspective to climate-security links by adopting an ANT-based research strategy – a sociology of translations (see Chapter 4). This was used to approach climate-security links as a result of webs of relations formed by the work of actors to connect climate and specific conflicts through *translations*. *Actors* were defined as anything that appears to make a difference. *Epistemic resources* were considered to be inputs to the generation and circulation of knowledge integrated into the translations. The result of the translations were understood to be *actor-networks* – an enabling or constraining web of relations, which might include *black boxes* – unchallenged components.

Chapters 5 to 7 described how actors connected climate and the Darfur and Syria conflicts. The chapters focused on controversies about climate-conflict links to reveal details of the work of actors, and the study sought to avoid *a priori* assumptions about actors and knowledge resources. The ANT-based approach was selected because it enabled considering the concrete activities to merge two previously disconnected things – climate and security – due to its open-endedness about which actors are relevant, and because of its sensitivity to the role of knowledge in social phenomena. This allowed describing the making of climate-conflict links from four perspectives: (a) the translations by actors to connect climate and conflicts; (b) the knowledge mobilized for those translations; (c) the marginalizations and unintended consequences reproduced through the formation of the nexus;

and (d) the patterns of agency and influence within the resulting climate-conflict actor-networks.

Darfur and Syria were chosen due to their paradigmatic role in the climate-security debate. The empirical Chapters 5 to 7 described how climate-conflict links were debated in context of these two wars. The chapters outlined the conflict background and the evolution of the climate-conflict debates. For both, documents contributing to the debates were analyzed to identify controversies, which provided a structure for the empirical descriptions. After this, the translations by actors to settle each controversy were described, followed by a mapping of the knowledge resources involved in the translations. These two steps provided the basis for describing (a) the role of knowledge in the debate, and (b) the modes of ordering of the actor-networks, including how changing modalities of translations and knowledge resources shift patterns of agency and influence within the actor-networks.

Key findings and resulting challenges

Based on this approach, this book described how climate-conflict debates evolved as a semi-codified process encompassing initial narratives, technical reviews and the IPCC's inscriptions-based synthesis of evidence. Within these debates, facts about climate-conflict links were considered within contributions inscribed in documents and other presentations, and efforts were made to settle them through synthesis and review.

The complex role of knowledge

The webs of relations enabling the two climate-conflict debates often involved limited knowledge resources and focused on narratives of linear causal mechanisms, and scientific resources tended to be mobilized as a reaction to initial narratives – not as drivers of the debates. Knowledge resources were often diverse and fragmented, and lines between science and policymaking were regularly blurred. Second, the imbalances between the knowledge basis of nonhuman and human aspects of the debates illustrated key knowledge gaps, leading to marginalization and stereotypical depiction of local populations and possible confirmation biases. And third, the debates involve patterns of influence that changed with the predominant knowledge resources and modalities of translation. Against public expectations and declarations of science-drivenness of environmental issues, the sciences did not, at least in the context of the two conflicts, drive the development of climate-security nexuses. Although the initial narratives of climate-conflict links involved some references to scientific studies, studies devoted to climate-conflict links appeared as responses to the components of the already-existing narratives. Thus, the narratives influenced research by establishing the linear causal mechanisms as a subject of research, and scientific work reacted to that structure.

Thus, while environmental policy debates are characterized by declarations about the importance of science, nexuses might develop predominantly in a much less linear fashion: through compelling narratives, storylines, and assumptions of causal relations, which initially require only a general correspondence with scientific understandings and might be later consolidated by research. This resembles Haas's observation that the formation of linkages involves uncertainty and ambiguity, to which the actors often respond with simplifications, questionable precedents, and guesswork. Haas has highlighted:

> "Linking issues is a fallible man's way of marshalling what knowledge he has in order to attain his goals" and that often "causal sequences are assumed or guessed at rather than studied fully".
>
> (Haas 1980, 378)

As outlined in Chapter 4, the development of facts is a collective endeavor (Latour 1987). Actors do not act or marshal knowledge without relationships (Callon 1986) and "nobody acts alone" (Mol 2010, 256). Even Haas's "fallible man" articulating the initial narratives must "stand on the shoulders of giants" by drawing on general understandings of climate change, statements by scholars and practitioners, and scientific studies. As outlined in Chapter 4, the criteria for agency in this book is the making of difference (Latour 2005; Mol 2010), and technical inscriptions make a difference (Hogle 2008; Aradau 2010). In this manner, the initial narratives of climate-conflict links were consolidated or criticized on the basis of inscriptions that generated detailed knowledge about the conflicts, and modalities of review of evidence inscribed in international law established filters that qualify or disqualify previous parts of the debate as evidence. Thus, while a lot of "guessing" of causal relations might seem to be happening within the climate-conflict debates, a large actor-network of inscription devices enabled inscribing detailed knowledge on those relations and settling debates about the "facts". These are what Aradau (2010) describes as the infrastructures that ensure the continuity and circulation of social phenomena. Without these inscriptions, there would be little knowledge to marshal or causal sequences to guess.

However, as of 2022, that debate has not created a *black box* of consensus around climate-conflict links, even though for both Darfur and Syria, IPCC has ruled against such links. The debates and the accompanying institutionalization continue regardless of pertinent public controversies, with no consensus in sight (see IPCC 2022). As highlighted in Chapter 8, these contradictory dynamics between expectations of science-drivenness, lack of evidence and accelerating securitization relate to the complex role of knowledge. This book indicated that intensification of the climate-security discourse in light of contested evidence is possible because science-policy relations within the climate-conflict issue do not correspond to an idealized model of two separate domains where objective facts precede normative policy decisions. The relationship can rather be described as co-productive (Jasanoff 2004; Irwin 2008). The co-productionist dynamic is illustrated by the dialogue between policy and scientific statements, the involvement of government institutions in the generation of knowledge resources, as well as the roles of some actors in publishing both policy statements and scientific outputs.

The Darfur and Syria debates further illustrated that (a) the articulations of linear causal mechanisms do not involve a fully studied information basis for all components of those mechanisms; (b) throughout the debate, the evidence basis for the mechanisms is not equally distributed across its components, and (c) sometimes the evidence is limited or questionable. This points toward a fragmented character of the knowledge basis. The stages of translations also indicated how both climate-conflict debates advance as a mutually influencing dialogue between initial problematizations of the links and scientific inscriptions. Furthermore, arrangements that enable the production of knowledge related to climate-conflict links were closely intertwined with national governments, who established data processing institutions and steered the work of the IPCC. Finally, some actors operated on both sides of the assumed science-policy divide, simultaneously working for advocacy organizations and publishing in academic journals.

To some extent these nuances are part of a perfectly normal process: policymakers advance arguments, and scientific studies scrutinize the truthfulness of those arguments. However, this book has also highlighted the blurred lines between policy and science work, the inability of scientific studies to consider the full scope of the climate-conflict narratives, as well as the risk of confirmation bias. These mean that the scientific scrutiny is likely to be imperfect, closely intertwined with the policy discussion and possibly misdirected by the compelling narratives of the linear causal mechanisms.

So when it comes to the role of knowledge in policymaking, this book illustrates several challenging aspects. First, the power of narratives within the climate-conflict debates seems incompatible with science-drivenness. The debates begin with coherent narratives accompanied by limited evidence across the components of those narratives. This shows a potentially problematic trend in international politics where actors seek good "stories" instead of solid evidence-based relations. For example, in the UNFCCC Talanoa Dialogue, an effort to enhance ambition of climate action, participants were invited to tell "stories" about their actions. Stories are, of course, important sources of inspiration and lessons. However, authors associated with prospect theory have observed the dangerous ease and attractiveness of coherence, which leads to heuristic biases and bad decision-making:

> The explanatory stories that people find compelling are simple; are concrete rather than abstract; . . . and focus on a few striking events that happened rather than on the countless events that failed to happen (Kahneman 2012, 199) . . . Paradoxically, it is easier to construct a coherent story when you know little, when there are fewer pieces to fit into the puzzle.
>
> (Ibid., 201)

Second, policies not based on solid factual evidence are likely to be less effective or counterproductive. Haas worried that inaccurate shared meanings in regime formation will lead to ineffective solutions and to "disappointment and suffering" (1990, 240). Chapters 5 and 6 highlighted, for example, how the climate-conflict links were evoked by the governments in both Khartoum and Damascus as excuses for their own policies and as a rhetorical tool to distract from their responsibility. Others have highlighted that climate-security risks can incentivize unproductive militarization, "climate fatigue" and ineffective climate change mitigation (see, i.e., Barnett 2001; Brown et al. 2007; Mayer 2012; McDonald 2013).

Third, the patterns of marginalization and stereotypical presentation are problematic from a normative standpoint, since they silence the people impacted by the conflicts, and turn them into avatars of Malthusian scarcity imageries. They are also problematic from a knowledge standpoint, because limited knowledge of local circumstances is likely to reduce the likelihood of success of any policies that might be developed to reduce conflict risks or to respond to emerging crises.

Fourth, the regularly limited evidence basis of climate-conflict narratives provides targets for climate skeptics, who can use the knowledge gaps to discredit climate policies more generally. Latour has described that by arguing that the world has full certainty and needs to stop debating and start acting, activists expose themselves to criticisms because climate skeptics can easily find information to call such full certainty into question (2018, 23–8) – but as any IPCC report demonstrates, climate sciences never operate with full certainty but with probabilities. This is certainly true of the linear causal mechanisms connecting climate and the Darfur and Syria conflicts, which suffer from several information gaps.

Fifth, climate activists urge the world to "unite behind the science", the Paris Agreement called for climate policies to be based on the best science, and the COVID-19 pandemic illustrated the importance of science-driven policies. However, this book also demonstrated how actors supportive of tackling environmental issues might publicly emphasize the importance of science, but in practice base their claims about climate-conflict links on limited scientific evidence. In addition, calls by other researchers for a better knowledge basis for climate-security links have triggered confrontations and limited mutual understandings (see Chapters 5 and 6, as well as Ide 2018), indicating that defending the coherence of the climate-conflict narratives sometimes takes precedence over constructive exchanged to enhance the factual basis of climate-conflict debates. Thus, if even the advocates of climate action are not working to ensure the science-drivenness of their arguments, and avoid constructive engagement with critics to strengthen the facts, then are climate-conflict narratives evolving toward "islands of consensus" on which the converted live parallel epistemic lives? If yes, how can science-drivenness of international policymaking be enhanced at all?

However, regardless of these challenges, this book has highlighted that power and influence do not reside in monolithic powerful actors and institutions, but can change as associations, modalities of translation and knowledge resources change. Thus, positive change is possible through incremental changes achieved by building of new associations. On that note, the next section considers some possible channels for addressing the challenges outlined in this one.

How to better integrate good knowledge into policymaking?

Clearly, there is a need to consider how the integration of high-quality knowledge into global policy debates can be enhanced. This is due to both the need to enhance the quality of policies and avoid unintended normative consequences, such as the marginalization of local populations. Given the co-productionist character of science-policy relations within the climate-conflict debates, it seems counter-productive to pursue solutions based on the idealized science-policy dichotomies discussed in Chapter 4. Also, setting up a new IPCC or developing a novel "vast machine" for considering the entire range of information relevant to climate-security links does not seem realistic. Instead, improvements could be achieved, for example, through incremental change in the practices and modalities of exchange within global institutions. As discussed in Chapter 8 and earlier, change is possible through new information, new associations and new modalities of interaction.

One basic option would be, as suggested by Ide (2018), to convene mediated meetings between climate-security scholars to discuss areas of divergence and convergence. Such meetings could highlight uncertainties, areas that require further research, and recommendations for policymakers. They could be organized under the auspices of institutions engaged in climate-security work, such as the UNCSM, the IPCC or the CCS to enhance their direct policy-relevance. On the basis of the Darfur and Syria, it seems that a particularly important subject would be to improve the evidence basis for the human aspects of the climate-conflict debates.

To further address the imbalance of evidence between natural and human aspects and its marginalizing effects, it would be essential for climate-security debates to be better informed by socioeconomic and ethnographic information about local circumstances in fragile contexts, and, where possible, by quality information from journalistic and non-governmental sources operating at the frontlines of crisis regions. Such information would not only help understand local circumstances but also provide more immediate information about fast-moving conflict situations. As a solution, for example, the IPCC could strengthen its ongoing efforts to better consider socio-economic literature, and, instead of categorically excluding information from journalistic and non-governmental sources from its reviews, develop modalities to review and critically consider such information, and, if deemed reliable, integrate it in its reports as a separate category from peer-reviewed studies. Such steps would help integrate socio-economic and more immediate information in global policy debates, and thus reduce the imbalance between natural and human questions within climate-security debates. Obviously, they would require consent from governments, but such steps would greatly enhance the science-drivenness of climate-conflict debates.

Furthermore, the UNCSM, a nascent arrangement to enhance the knowledge basis of climate-security debates, could evolve into a forum for considering all aspects, in particular human components, of climate-security links by drawing on a wide range of different types of research, including quantitative and qualitative studies and ethnographic research, as well as immediate local information on potential conflict dynamics. The Mechanism could also provide a fact-checking function in relation to climate-conflict debates.

Further areas for research

In focusing on Darfur and Syria, this study covered only a small portion of the climate-security debate. Future research could delve into other conflicts that have been described as climate wars, including Lake Chad (Nagarajan et al. 2018; Vivekananda et al. 2019), Somalia (UNSC 2011) and Yemen (Douglas 2016). Many other areas, including the Arctic (IPCC 2022, 7–63), Sub-Saharan Africa (Burke et al. 2009) and Bangladesh (Day and Caus 2020), have been presented as examples of climate-conflict links. These provide opportunities for further study of the efforts to connect climate and conflicts, thus broadening the empirical understanding of the formation of the climate-security nexus, and verifying, challenging or complementing this study. Research could also focus on how climate-security risks are approached as so-called "compound risks" – a novel way to complex risks in global policy conversations (Rüettinger et al. 2015; UNCSM 2020; Zelli et al. 2020).

While climate-security links are controversial, this book has highlighted how the debate is evolving toward further institutionalization. This includes the UNCSM, the Berlin Climate Security Conference (Adelphi and PIK 2020(a), 2020(b)), the Group of Friends of Climate and Security, the NATO Center of Excellence for Climate and Security, as well as work of several governments NGOs and other initiatives (Krampe and Mobjörk 2018). Studies could investigate this institutionalization and policy implementation related to climate-security links and complement existing research on regime formation (Keohane 1984; Haas 1980, 1990) in particular by considering how nexus governance emerges under controversies.

Finally, following Mayer (2012) and Rothe (2017), research could delve into patterns of influence within actor-networks of environmental policymaking. Mayer (2012) has described how global climate-security assemblages and discourses tend to reinforce national interests, prioritize adaptation and legitimate interventionism. Rothe (2017) outlined how the assemblage of visual technologies to understand climate change might promote paternalistic control, favor well-resources actors and silence discursive struggles. This book indicated how the two climate-conflict actor-network are driven by linear causal mechanisms, involve limited information on the human aspects, and feature patterns of influence that marginalize contributions that do not conform with the requirements of institutionalized science. Further focus on such patterns in other contexts would help understand the micro-expressions of power within international policy debates.

Final remarks

This book has considered how actors connected climate change and the Darfur and Syria conflicts, the knowledge resources they mobilized, and the consequences of building such connections. This book did not provide a full account of the climate-security debates, or determine whether or why conflicts and climate are connected (although some positioning in that regard was inevitable in light of the empirical findings). Rather, the main objective was to understand the role of knowledge in all its diversity beyond traditional assumptions about science-policy relations. This book also did not question the scientific consensus on or the reality and seriousness of anthropogenic climate change – those facts are well-established and the urgent need for climate action is obvious.

However, this book did highlight the importance of caution when articulating links between climate and complex social phenomena. Such links are often articulated with limited evidence and with the power of coherent narratives. While the vast machine of the climate sciences provides a clear overall picture of the global climate system and its ongoing and dramatic change, it has yet to shed light to a similar degree on the complex interactions between groups of humans hit by climate impacts. The descriptions in this book have aimed to highlight some of the epistemic blind spots in international policy debates, and to point toward potential solutions, as well as toward ways in which such solutions could be productively articulated in light of the co-productive nature of science-policy relations.

This book also sought to complement existing climate-security research, which has focused on climate-conflict correlations, intervening variables and climate-security discourses by providing empirical descriptions of the knowledge-related processes in linking climate and conflicts in context of the two paradigmatic climate wars. It contributed to study of issues-linkages and nexuses in IR research by considering how actors connect two previously disconnected hegemonial governance areas of climate and security into a nexus, illustrating how the climate-security nexus is being built, among other processes, by connecting climate change and two deadly conflicts. It also added to existing research that aims to understand the role of production of knowledge in the formation of nexuses of governance areas in international politics, highlighting the co-productive character of knowledge in nexus formation. While existing IR research has touched on such questions, they are yet to emerge as a broader research agenda.

With this in mind, possibly the contradictions between accelerating securitization of climate change and limited evidence relate to tendencies to prioritize coherent stories over carefully selected evidence. As illustrated in this book, the world is full of unexpected connections and incoherent turns – which do not easily make for good stories, and which require extensive work to be understood and described. If compelling narratives increasingly take priority over solid research and evidence, then the world will have lost the aspiration of the Paris Agreement toward science-driven policymaking, and the securitization of climate change may continue accelerating, no matter how controversial the facts.

Literature

Adelphi and Postdam Institute for Climate Impact Research (PIK) 2020(a): *Summary – Berlin climate and security conference, Part I.* At: https://berlin-climate-security-conference.de/sites/berlin-climate-security-conference.de/files/documents/summary_bcsc_2020_part_i.pdf (accessed 30 May 2021).

Adelphi GmbH and Postdam Institute for Climate Impact Research (PIK) 2020(b): *Summary – Berlin climate and security conference, Part II.* At: https://berlin-climate-security-conference.de/sites/berlin-climate-security-conference.de/files/documents/bcsc_2020_part_ii_online_summary_0.pdf (accessed 30 May 2021).

Aradau, C. 2010: *Security that matters: Critical infrastructure and objects of protection.* In: Security Dialogue, 41(5), 491–514.

Barnett, J. 2001: *Climate Change and Security.* Tyndall Center for Climate Change Research, Manchester.

Brown, O.; Hammill, A. and McLeman, R. 2007: *Climate change as the 'new' security threat: Implications for Africa.* In: International Affairs, 83(6), 1141–54.

Burke, M.B.; Miguel, E.; Satyanath, S.; Dykema, J.A. and Lobell, D.B. 2009: *Warming increases the risk of civil war in Africa.* In: PNAS, 106(49), 20670–4.

Callon, M. 1986: *Some elements of a sociology of translation: Domestication of the scallops and the fishermen of St. Brieuc Bay.* In: Law, J. (Ed.): Power, Action and Belief: A New Sociology of Knowledge? Routledge, London, 196–223.

Day, A. and Caus, J. 2020: *Conflict Prevention in an Era of Climate Change: Adapting the UN to Climate-Security Risks.* United Nations University, New York.

Douglas, C. 2016: *A Storm Without Rain: Yemen, Water, Climate Change and Conflict.* 3 August. At: https://climateandsecurity.org/2016/08/a-storm-without-rain-yemen-water-climate-change-and-conflict/ (accessed 24 July 2022).

GWR 2019: *First Climate Change War.* At: www.guinnessworldrecords.com/world-records/first-climate-change-war (accessed 7 September 2021).

Haas, E.B. 1980: *Why collaborate?: Issue-linkage and international regimes.* In: World Politics, 32(3), 357–405.

Haas, E.B. 1990: *Reason and change in international life: Justifying a hypothesis.* In: Journal of International Affairs, 44(1), 209–40.

Hackett, E.J.; Amsterdamska, O.; Lynch, M. and Wajcman, J. (Eds.) 2008: *The Handbook of Science and Technology Studies.* 3rd Edition. MIT Press, Cambridge, MA.

Hogle, L.F. 2008: *Emerging Medical Technologies.* In: Hackett et al. 2008, 841–74.

Ide, T. 2018: *Climate war in the Middle East? Drought, the Syrian civil war and the state of climate-conflict research.* In: Current Climate Change Reports, 4(4), 347–54.

IPCC 2022: (H.-O. Pörtner, D.C. Roberts, M. Tignor, E.S. Poloczanska, K. Mintenbeck, A. Alegría, M. Craig, S. Langsdorf, S. Löschke, V. Möller, A. Okem, B. Rama (eds.)): *Climate Change 2022: Impacts, Adaptation, and Vulnerability. Contribution of Working Group II to the Sixth Assessment Report of the IPCC.* Cambridge University Press. In Press. Available at: www.ipcc.ch/report/sixth-assessment-report-working-group-ii/ (accessed 19 June 2022).

Irwin, A. 2008: *STS perspectives on scientific governance.* In: Hackett et al. 2008, 583–608.

Jasanoff, S. (Ed.) 2004: *States of Knowledge – The Co-Production of Science and Social Order.* Routledge, Oxon.

Kahneman, D. 2012: *Thinking, Fast and Slow.* Penguin, London.

Keohane, R. 1984: *After Hegemony: Cooperation and Discord in the World Political Economy.* Princeton University Press, Princeton, NJ.

Krampe, F. and Mobjörk, M. 2018: *Responding to climate-related security risks: Reviewing regional organizations in Asia and Africa.* In: Current Climate Change Reports, 4, 330–7.

Latour, B. 1987: *Science in Action – How to Follow Scientists and Engineers Through Society.* Harvard University Press, Cambridge, MA.

Latour, B. 2005: *Reassembling the Social – an Introduction to Actor-Network Theory.* Oxford University Press, Oxford.

Latour, B. 2018: *Facing Gaia: Eight Lectures on the New Climatic Regime*. Polity Press, Cambridge.

Mayer, M. 2012: *Chaotic climate change and security*. In: International Political Sociology, 6, 165–85.

McDonald, M. 2013: *Discourses of climate security*. In: Political Geography, 33, 42–51.

Mol, A. 2010: *Actor-network theory: Sensitive terms and enduring tensions*. In: Kölner Zeitschrift für Soziologie und Sozialpsychologie, Sonderheft, 50, 253–69.

Nagarajan, C.; Pohl, B.; Rüttinger, L.; Sylvestre, F.; Vivekananda, J.; Wall, M. and Wolfmaier, S. 2018: *Climate-Fragility Profile: Lake Chad Basin*. Adelphi, Berlin. At: www. adelphi.de/en/system/files/mediathek/bilder/Lake%20Chad%20Climate-Fragility%20 Profile%20-%20adelphi_0.pdf (accessed 21 November 2019).

Rothe, D. 2017: *Seeing like a satellite: Remote sensing and the ontological politics of environmental security*. In: Security Dialogue, 48(4), 334–53.

Rüttinger, L.; Smith, D.; Stand, G.; Tänzler, D. and Vivekananda, J. 2015: *A New Climate for Peace – Taking Action on Climate and Fragility Risks*. An Independent Report Commissioned by the G7 Members and Prepared by Adelphi, International Alert, Woodrow Wilson International Center for Scholars, and the EU Institute for Security Studies, Berlin.

United Nations Climate Security Mechanism (UNCSM) 2020: *Toolbox: Briefing Note*. New York. At: https://dppa.un.org/sites/default/files/csm_toolbox-1-briefing_note.pdf (accessed 17 July 2022).

UNSC 2011: *Record of the 6587th Meeting*. Document S/PV.6587. At: https://undocs. org/en/S/PV.6587 and https://undocs.org/en/S/PV.6587(Resumption1) (accessed 20 December 2020).

UNSC 2021: *Maintenance of International Peace and Security: Climate and Security – Security Council Open VTC*. 23 February. At: http://webtv.un.org/search/maintenance-of-international-peace-and-security-climate-and-security-security-council-open-vtc/623468 6966001/?term=&lan=english&cat=Security%20Council&sort=date&page=5 (accessed 29 May 2021).

Vivekananda, J.; Wall, M.; Sylvestre, F. and Nagarajan, C. 2019: *Shoring Up Stability: Addressing Climate and Fragility Risks in the Lake Chad Region*. Adelphi, Berlin. At: www.adelphi.de/en/publication/shoring-stability (accessed 23 November 2019).

Zelli, F.; Bäckstrand, K.; Nasiritousi, N.; Skovgaard, J. and Widerberg, O. 2020: *Governing the Climate-Energy Nexus: Institutional Complexity and Its Challenges to Effectiveness and Legitimacy*. Cambridge University Press, Cambridge.

Index

Note: Page numbers in **bold** indicate a table and page numbers in *italics* indicate a figure on the corresponding page.

Printed in the United States
by Baker & Taylor Publisher Services

Printed in the United States
by Baker & Taylor Publisher Services